AF861297

SPECIAL TRIGONOMETRIC SERIES IN k-DIMENSIONS

Stephen Wainger*

The purpose of this paper is to make a study of special classes of trigonometric series in k-dimensions. These series are analogues of one-dimensional series which have shed light on problems in Fourier analysis in one dimension. This study is carried out in parts I-III.

In part IV, we give applications to various problems in analysis. The applications and their relations to known theorems are discussed in detail in part IV. Here, we merely state the applications:

i) We show that there exists a function $f(x)$ defined on the k-dimensional torus such that f is in Lip $k/2$ and such that the Fourier series of f does not converge absolutely.

ii) We give a natural definition of fractional integration of trigonometric series in any number of dimensions, and we prove the usual elementary facts about this fractional integral.

iii) We consider the set of numbers

$$\gamma(\alpha,\beta) = \begin{cases} |n|^{-\beta}\exp(i|n|^{\alpha}), & n \neq 0, \\ 0 & n = 0. \end{cases}$$

(n runs over the lattice points of k-dimensional space.) We assume $\beta < \frac{1}{2}k\alpha$, and $0 < \alpha < 1$. Then $\gamma(\alpha,\beta)$ is a multiplier on L^p if $2k\alpha(k\alpha + 2\beta)^{-1} < p < 2k\alpha(k\alpha - 2\beta)^{-1}$. If $p < 2k\alpha(k\alpha + 2\beta)^{-1}$ or if $p > 2k\alpha(k\alpha - 2\beta)^{-1}$, $\gamma(\alpha,\beta)$ is not a multiplier on L^p.

*Supported in part by NSF Grant GP 1645.

iv) We give an example of a function $f(x)$ such that

$$\int |f(x)| \{\log(1 + |f(x)|)\}^{1-\varepsilon} dx < \infty$$

for every $\varepsilon > 0$ and such that the Riesz transforms of f are not integrable.

We study three basic types of series. The first type is discussed in part I. The simplest result here is

1) $$\sum{}' |n|^{-\gamma} \exp(2\pi i n \cdot x) \simeq K_\gamma |x|^{\gamma - k}$$

as $x \to 0$, for $0 < \gamma < k$.

To study fractional integration and Riesz transforms, we must introduce slowly varying factors. Thus, for example we obtain

$$\sum{}' (|n|^\gamma \log|n|)^{-1} \exp(2\pi i n \cdot x) \simeq K_\gamma |x|^{\gamma - k} (\log(1/|x|))^{-1},$$

$0 < \gamma < k$. For our applications, we shall also need to examine the extreme cases $\gamma = 0$ and $\gamma = k$. In these extreme cases we obtain slightly different results.

Finally our applications to Riesz transforms require us to study series of the form

$$\sum{}' W(n')(|n|^\gamma \log|n|)^{-1} \exp(2\pi i n \cdot x);$$

$n' = n/|n|$, and W is a homogeneous function of degree 0.

Thus we see that the theorems of part I must contain many special cases, and they cannot be expressed concisely.

The simplest theorem of part III states that

2) $$\sum{}' |n|^{-b} \exp(i|n| \log|n| + 2\pi i n \cdot x)$$

represents a continuous function if $b > \frac{1}{2}k$. Hence 2) is the Fourier series of a continuous function as soon as the coefficients are in ℓ^2. In order to consider derivatives of such series, we need to consider

$$\sum{}' |n|^{-b} Y_\ell(n') \exp(i|n| \log|n| + 2\pi i n \cdot x);$$

$Y_\ell(n')$ is a (homogeneous) spherical harmonic of degree ℓ. For our applications to the absolute convergence of Fourier series, we must introduce various logarithmic factors. Thus we are led to examine series of the form

$$\sum{}' (|n|^b \log^c |n|)^{-1} Y_\ell(n') \exp\{i|n|\log^d|n| + 2\pi i n \cdot x\}.$$

In part II, we discuss functions represented by a series of the form

$$\sum{}' |n|^{-b} \exp(i|n|^a + 2\pi i n \cdot x); \tag{3}$$

$0<a<1$; if $0<b\leq k - \frac{1}{2}ka$. We show 3) is asymptotic to

$$K_a(x)\, |x|^{(\frac{1}{2}ka + b - k)/(1-a)}$$

as $x \to 0$. $K_a(x)$ is a bounded oscillatory function.

The series treated in parts I and III are, in some sense, extreme cases of those studied in part II. For example, the series of part I do not represent a bounded function unless the series is nearly absolutely convergent. On the other hand, the series of part III represent a bounded function if the coefficients are little better than ℓ^2. Now as a ranges from 0 to 1, the coefficients of the series of part II must nearly belong to the various ℓ^p spaces, $1 < p < 2$, in order that the series represent a bounded function.

There is one unifying feature throughout parts I-III. This is the use of the Poisson summation formula to gain information about trigonometric series from knowledge of trigonometric integrals. Unfortunately, the Poisson summation formula as it stands did not apply to all the cases in which we wished to use it. This led us to consider not the Fourier transform of a function f, but rather the Fourier transform of a slightly modified function,

$$f_1(x) = \psi(|x|)f(x)\exp(-\varepsilon|x|).$$

$\psi(t)$ is in $C^\infty(-\infty,\infty)$, $\psi(t) = 0$ for $t\leq\frac{1}{2}$, and $\psi(t) = 1$ for $t \geq 1$.

This method has also been used by Bochner (see [4]) to study the behavior of the series

$$\lim_{\varepsilon\to 0+} \sum{}' |n|^{-\gamma} W(n')\exp(-\varepsilon n + 2\pi i n\cdot x)$$

as an analytic function of γ.

Once our problems concerning the behavior of trigonometric series are changed to problems concerning trigonometric integrals, we use two different methods. The simpler of these conceptually is integration by parts, which we use in part I. In parts II and III we employ the method of steepest descents.

The work in this paper is based on my PhD thesis at the University of Chicago written under the supervision of Professor E. M. Stein. I would like to thank Professor Stein for his help, encouragement, and patience. I would also like to thank Professors A. P. Calderón and A. Zygmund for their aid. Finally, I would like to thank Professor I. I. Hirschman for his many suggestions on the organization and exposition of this paper, and for the correction of several errors.

NOTATION

We shall now introduce notation which we shall use throughout. Occasionally we shall repeat certain notation in the body of our paper for the sake of clarity. E_k will denote k-dimensional Euclidean space. x and y will be vectors in E_k with components $x_1, \ldots, x_k$ and $y_1, \ldots, y_k$ respectively. R will be equal to $|x| = (x_1^2 + \cdots + x_k^2)^{1/2}$, and $x \cdot y = x_1 y_1 + \cdots + x_k y_k$. S_{k-1} will be the unit sphere in E_k, that is, S_{k-1} will consist of those points x for which $R = 1$. For $x \neq 0$, we let $x' = x/|x|$. Hence x' is in S_{k-1}. $Y_{n,\ell}(x')$ will be a (homogeneous) spherical harmonic of degree n. See [6] chapter 6, p. 31, for the definition of spherical harmonic. The Fourier transform of a function f(x) in $L^1(E_k)$ will be denoted by $f^\wedge(x)$ and defined by the relationship

$$f^\wedge(x) = \int_{E_k} f(y)\exp(-2\pi i x \cdot y)\,dy.$$

The inverse Fourier transform of f will be designated by $f^\vee$ and defined by

$$f^\vee(x) = \int_{E_k} f(y)\exp(2\pi i x \cdot y)\,dx.$$

More generally, if μ is a finite measure, we set

$$\mu^{\wedge}(x) = \int_{E_k} \exp(-2\pi i x \cdot y)\, d\mu(y).$$

We call $\mu^{\wedge}$ the Fourier Stieltjes transform of the measure μ. Also we let

$$\mu^{\vee}(x) = \int_{E_k} \exp(2\pi i x \cdot y)\, d\mu(y).$$

$\mu^{\vee}$ is the inverse Fourier Stieltjes transform of the measure μ.

$\sum a(n)$ will mean that we sum the numbers $a(n)$ over the lattice points, n, in E_k. (We shall only consider sums which are absolutely convergent, so there will be no ambiguity as to the method of ordering the lattice points.) $\sum' a(n)$ is the sum over those lattice points such that $a(n)$ is finite.

Let $f(x)$ and $g(x)$ be two functions. We shall write $f(x) \simeq g(x)$ as $x \to a$ (or if there is no danger of confusion merely $f(x) \simeq g(x)$) if $f(x)[g(x)]^{-1} \to 1$ as $x \to a$. If $f(x)[g(x)]^{-1}$ is bounded, we shall write $f(x) = O(g(x))$; if $f(x)[g(x)]^{-1} \to 0$, we shall say that $f(x) = o(g(x))$.

A function $f(x)$ will be called periodic if $f(x_1, \ldots, x_{j-1}, x_j+1, x_{j+1}, \ldots, x_k) = f(x_1, \ldots, x_{j-1}, x_j, x_{j+1}, \ldots, x_k)$ for all $j = 1, 2, \ldots k$. T^k will be the set of points x in E_k such that $-1/2 \le x_j \le 1/2$, $j = 1, \ldots, k$. A periodic function is completely determined by its values on T^k. If $f(x)$ is integrable on T^k, we define

$$c(n) = c(n, f) = \int_{T^k} f(x) \exp(-2\pi i n \cdot x)\, dx.$$

n runs over the lattice points of E_k, and we say that the numbers $c(n)$ are the Fourier coefficients of f.

Unless otherwise stated, differentiation is in the ordinary sense. C^{∞} will be the space of infinitely differentiable functions, and C_0^{∞} will denote the space of infinitely differentiable functions with compact support. Let s be a k-dimensional vector with integral components, i.e., $s = (s_1, s_2, \ldots s_k)$ with the s_j integers. Then

$$D^s f = \frac{\partial^{s_1 \cdots s_k}}{\partial x_1^{s_1} \cdots x_k^{s_k}} f.$$

Also for any k-dimensional vector y, $(y)^s = y_1^{s_1} \cdots y_k^{s_k}$. By $\|s\|$ we mean $s_1 + s_2 + \ldots + s_k$. We denote the Laplacean of f by Δf. $\Delta^2 f = \Delta(\Delta f)$ and for any integer $r \geq 2$, $\Delta^r f = \Delta(\Delta^{r-1} f)$.

We shall reserve the letter ψ to denote a function having the following properties:

(i) $\psi(t)$ is in C^∞ $(-\infty, \infty)$.

(ii) $\psi(t) = 0$ for $t \leq 1/2$.

(iii) $\psi(t) = 1$ for $t \geq 1$.

(iv) $0 \leq \psi(t) \leq 1$ for $-\infty < t < \infty$.

Let a be a real number. Then by $[a]$ we mean the largest integer equal to or less than a. If $0 \leq a < 1$, we say that f is in Lip a if

$$|f(x+h) - f(x)| = O(|h|^a).$$

If $a \geq 1$, we say that f is in Lip a if the following two conditions are satisfied:

(i) The partial derivatives of f of order m exist and are continuous for $0 \leq m \leq [a]$.

(ii) $D^s f$ is in Lip $\{a - [a]\}$ for $\|s\| = [a]$.

(Our definition of Lip 1 is slightly stronger than the usual definition.)

We shall use Γ to denote the Γ function.

$$(a)_n = \Gamma(a+n)/\Gamma(a) = a(a+1)\ldots(a+n-1).$$

Let z be a complex number. Then $\mathcal{R}e(z)$ will denote the real part of z, and $\mathcal{I}m(z)$ will be the imaginary part of z. Finally, integers in square brackets refer to the appropriate reference in the bibliography.

PART I

We begin with two definitions.

DEF. 1. We shall say that a function b of a non-negative real variable t is slowly varying if b(t) satisfies the following three conditions.

i) For any $\delta > 0$, there is a $t_0(\delta)$ such that $t^{\delta}|b(t)|$ is increasing for $t > t_0(\delta)$.

ii) For any $\delta > 0$, there is a $t_1(\delta)$ such that $t^{-\delta}|b(t)|$ is decreasing for $t > t_1(\delta)$.

iii) b(t) is differentiable for $t \geq 0$. (At $t = 0$ we mean differentiable from the right.) This definition is similar to that of [25], Vol. I, p. 186. We find it convenient however to allow b(t) to be negative. Also we are more precise about the smoothness of b(t).

Let b(t) be a slowly varying function which is infinitely differentiable in $[0, \infty)$. Let $h_0(t) = b(t)$, and define

$$h_j(t) = t\frac{d}{dt}h_{j-1}(t)$$

for $j = 1, 2, \ldots$. We use the $h_j(t)$ in the following definition.

DEF. 2. S is defined to be the subset of slowly varying functions b(t) possessing the following two properties:

i) b(t) is infinitely differentiable in $[0, \infty)$.

ii) The $h_j(t)$ associated to b(t) are slowly varying for all j.

We now give several common examples of functions in S.

i) $\log^a(t+2)$ (a either positive or negative)

ii) $\log \log^a (t + 100)$

iii) $\log^a(t+2) \log \log^b(t + 100)$.

In the appendix we give a large class of slowly varying functions which are in S.

As we mentioned in the introduction, our theorems about trigonometric series will be consequences of our results concerning trigonometric integrals. Therefore our first theorems concern integrals rather than series.

In Theorem 1, as throughout the paper, $\psi(t)$ is in $C^\infty(-\infty, \infty)$, and $\psi(t)$ has the following properties:

i) $\psi(t) = 0$ for $t \leq 1/2$.

ii) $\psi(t) = 1$ for $t \geq 1$.

iii) $0 \leq \psi(t) \leq 1$ for all t.

Also $Y_\ell(x')$ is a spherical harmonic of degree ℓ. (ℓ is a non-negative integer.) See [6], chapter 6, p. 31.

THEOREM 1. Let $b(t)$ be in S. Assume $-\infty < \gamma < k+\ell$, and also suppose $\gamma+\ell \neq 0, -2, -4, \ldots$ Define

$$F_\varepsilon(x) = \{\psi(|y|)|y|^{-\gamma} b(|y|)Y_\ell(y')\exp(-\varepsilon|y|)\}^{\vee}.$$

Then

i) $F(x) = \lim_{\varepsilon \to 0+} F_\varepsilon(x)$ exists for $x \neq 0$, and $F(x)$ is infinitely differentiable for $x \neq 0$.

ii) For any integer r, $|F_\varepsilon(x)| = O(R^{-r})$ uniformly in $\varepsilon \geq 0$, as $x \to \infty$.

At $x = 0$, we have

1.1 $$F(x) = i^{-\ell}\pi^{\gamma-k/2}\,\Gamma\left(\frac{1}{2}(k+\ell-\gamma)\right)\left(\Gamma\left(\frac{1}{2}(\ell+\gamma)\right)\right)^{-1} Y_\ell(x')b(1/R)R^{\gamma-k} + E(x).$$

1.2 $$|E(x)| = O(R^\ell) + o(R^{\gamma-k}\, b(1/R)) \quad \text{as } x \to 0.$$

Also, if $F(x)$ is in L^1, then $F_\varepsilon(x)$ is dominated by an L^1 function, and

$$F^\wedge(x) = \psi(|x|)|x|^{-\gamma}b(|x|)Y_\ell(x').$$

We now add several remarks to Theorem 1.

REMARK 0. The term $O(R^\ell)$ in equation 1.2 is needed in case $b(t) = 0$ for all sufficiently large values of t.

REMARK 1. $F(x)$ is in L^1 if $\gamma > 0$. If $\gamma = 0$, $F(x)$ is in L^1 if and only if $\int_1^\infty |b(t)| t^{-1} dt < \infty$. If $\gamma = k$, $F(x)$ is bounded if and only if $b(t)$ is bounded. (If $\gamma = k$, the hypothesis imply $\ell > 0$.)

REMARK 2. If $b(t) \equiv 1$ and if $0 < \gamma < k$, then $E(x)$ is infinitely differentiable, and any derivative of $E(x)$ is uniformly bounded.

REMARK 3. Assume

$$\int_{S_{k-1}} |Y_\ell(x')|^2 d\sigma = 1.$$

Then we can give estimates not depending on ℓ. That is, there exist positive numbers A and λ, not depending on ℓ, such that the following four conditions hold:

i) The $E(x)$ in equation 1.2 satisfies

$$|E(x)| \leq (1+\ell)^\lambda \{O(R)^\ell + o(1) R^{\gamma-k} |b(1/R)|\},$$

as $x \to 0$. The large O and little o do not depend on ℓ.

ii) $|D^s F(x)| \leq A(1+\ell)^\lambda$ for $x \neq 0$.

s is any k-tuple of integers, $s = (s_1, s_2, \ldots, s_k)$. A may depend upon s. The A is uniform for all x in any compact set which does not contain the origin.

iii) Let $|x| \geq 1$. Then $|F_\varepsilon(x)| \leq A(1+\ell)^\lambda R^{-r}$ for any integer r. A depends on r.

iv) If $F(x)$ is in L^1, then $\int_{E_k} |F(x)| dx \leq A(1+\ell)^\lambda$. (The condition that $F(x)$ be in L^1 is independent of ℓ by remark 1.) Remark 3 will be used in the proof of Theorem 2.

Proof: We shall break up the proof of Theorem 1 into several lemmas. In these lemmas A and λ will be positive constants not depending on ℓ.

LEMMA 1. Let $b(t)$ be slowly varying. Then

i) $b(t)$ is either non-positive or non-negative for sufficiently large t.

ii) $|b'(t)| = o(t^{-1}|b(t)|)$ as $t \to \infty$.

Proof:

i) There is a t_0 such that $t|b(t)|$ is increasing for $t \geq t_0$; therefore $tb(t)$ must be either non-positive or non-negative for $t \geq t_0$. Thus $b(t)$ must be either non-positive or non-negative for $t \geq t_0$. This proves i).

We turn now to ii). Let δ be any positive number. It suffices to exhibit a $t_2(\delta)$ such that

1.3 $$|b'(t)| < \delta t^{-1}|b(t)|$$

for $t \geq t_2(\delta)$. By part i) of this lemma, we may assume without loss of generality that $b(t) \geq 0$. Then by definition of slowly varying, there is a $t_0(\delta)$ such that $t^{\delta}b(t)$ is increasing for $t \geq t_0(\delta)$. Therefore, $d/dt\{t^{\delta}b(t)\} \geq 0$ for $t \geq t_0(\delta)$. Hence, for $t \geq t_0(\delta)$, $\delta t^{\delta-1} + t^{\delta}b'(t) \geq 0$, and we see

1.4 $$b'(t) \geq -\delta t^{-1}b(t); \qquad t \geq t_0(\delta).$$

Similarly, differentiation of $t^{-\delta}b(t)$ shows

1.5 $$b'(t) \leq \delta t^{-1}b(t) \qquad \text{for } t \geq t_1(\delta).$$

1.4 and 1.5 imply 1.3 with $t_2(\delta) = \max(t_0(\delta), t_1(\delta))$. This completes the proof of Lemma 1.

LEMMA 2. Let $b(t)$ be a slowly varying function. Let ξ_2 and ξ_3 be positive numbers with $\xi_2 < \xi_3$. Then

$$\max_{\xi_2/R \leq t \leq \xi_3/R} |b(t) - b(1/R)| = o(|b(1/R)|)$$

as $R \to 0$.

Proof: We may assume without loss of generality that $\xi_2 < 1 < \xi_3$. Suppose $\xi_2/R \leq t \leq 1/R$. Let $\delta > 0$. Then by Lemma 1, there is an R_1 such that

$|b'(t)| \leq \delta t^{-1}|b(t)|$ for $t \geq \xi_2/R_1$, and such that $t^{\frac{1}{2}}|b(t)|$ is increasing for $t \geq \xi_2/R_1$. Thus for $R < R_1$,

$$|b(t) - b(1/R)| = \left|\int_t^{1/R} b'(t)dt\right|$$

$$\leq \delta \int_{\xi_2/R}^{1/R} t^{-1}|b(t)|dt$$

$$\leq \delta \int_{\xi_2/R}^{1/R} t^{-3/2}\, t^{1/2}|b(t)|dt$$

$$\leq \delta R^{-1/2}\, |b(1/R)| \int_{\xi_2/R}^{1/R} t^{-3/2}\, dt$$

$$\leq 2\delta\xi_2^{-1/2}|b(1/R)|.$$

A similar argument shows that there exists an R_2 such that if $R < R_2$, and $1/R \leq t \leq \xi_3/R$,

$$|b(t) - b(1/R)| \leq 2\delta\xi_3^{1/2}|b(1/R)|.$$

Since δ is an arbitrary positive number, this completes the proof of Lemma 2.

LEMMA 3. Suppose $\alpha(x)$ is in $L^1(E_k)$. Also assume α only depends on $R = |x|$. Then

$$\{\alpha(x)Y_\ell(x')\}^\vee = 2\pi(-i)^\ell R^{(2-k)/2}Y_\ell(x')\cdot \int_0^\infty \bar{\alpha}(t)t^{k/2}\, J_{\frac{1}{2}(k-2)+\ell}(2\pi Rt)dt.$$

$\bar{\alpha}(t) = \alpha(|x|)$ if $t = |x|$.

Lemma 3 is proved in [4], p. 38, Theorem 2.6.1. Lemma 3 explains why we are interested in properties of Bessel functions. In the following lemmas μ and η will denote real numbers.

LEMMA 4. Suppose $b(t)$ is in S. Let ν be a positive integer. Assume $\eta \neq -1-\mu-\ell-2j$ for each integer j satisfying $0 \leq j \leq \nu-1$. Then

1.6 $$\int_0^\infty \psi(t)t^{-\eta} b(t)\exp(-\varepsilon t) J_{\mu+\ell}(2\pi tR)\,dt$$

$$= (\pi R)^{-\nu}\left(\tfrac{1}{2}(\eta+\ell+\mu+1)\right)_\nu \int_0^\infty \psi(t)t^{-\nu-\eta} b(t)\exp(-\varepsilon t)\cdot J_{\mu+\nu+\ell}(2\pi tR)\,dt$$

$$+ (\pi R)^{-\nu} \sum_{\substack{m+p+g+q=\nu \\ q<\nu}} d(m, p, g, q)\varepsilon^m$$

$$\cdot \int_0^\infty \psi^{(g)}(t)\, t^{-\eta-p-q} h_p(t)\exp(-\varepsilon t) J_{\mu+\nu+\ell}(2\pi tR)\,dt.$$

$$\left(\tfrac{1}{2}(\eta+\ell+\mu+1)\right)_\nu = 2^{-\nu}(\eta+\ell+\mu+1)(\eta+\ell+\mu+3)\ldots(\eta+\ell+\mu+2\nu-2).$$

The indices m, p, g, and q are integers. $|d(m, p, g, q)| \le A(1+\ell)^\nu$; A may depend upon ν. $\psi^g(t) = d^g/dt^g(\psi(t))$. $h_0(t) = b(t)$, and $h_p(t) = t\,d/dt\,(h_{p-1}(t))$ for $p \ge 1$.

Proof: To prove 1.6, we begin with the left-hand side and integrate by parts. There are ν "stages" of integration by parts, each consisting of $4^{\nu-1}$ integration by parts. At the jth stage we use the formula

$$J_{\mu+\ell+j-1}(2\pi tR) = (2\pi R)^{-1}(2\pi tR)^{-\mu-\ell-j} d/dt$$

$$\cdot \{(2\pi tR)^{\mu+\ell+j} J_{\mu+\ell+j}(2\pi tR)\}.$$

See [24], p. 45, sixteenth line from top of page. Also, at any stage we replace an $h'_p(t)$ by $t^{-1}h_{p+1}(t)$.

LEMMA 5. Let $t \ge 0$, and assume $\mu \ge -\frac{1}{2}$. Then for t complex and $|t| \le A$,

1.7 $$J_{\mu+\ell}(t) = 2^{-\mu-\ell}\, t^{\mu+\ell}\{\Gamma(\mu+\ell+1)\}^{-1} + E_5(t);$$

$$|E_5(t)| \le At^{\mu+\ell+1}\{\Gamma(\ell+1)\}^{-1}.$$

For real positive t

1.8 $$|J_{\mu+\ell}(t)| \leq A(\ell+1)^6 t^{-1/2}.$$

REMARK 1. 1.8 and 1.7 imply $|J_{\mu+\ell}(t)| \leq A(\ell+1)^6 t^{\mu+\ell}$ for all non-negative t.

REMARK 2. We shall not need 1.7 with complex t until we come to part II.

REMARK 3. The factor $(\ell+1)^6$ in 1.8 is by no means best possible. However, it suffices for our purposes.

Proof: 1.7 follows from [24], p. 45, sixteenth line from the top of the page. 1.8 is a consequence of various inequalities in [24]. For any $\theta < 1$, 1.8 follows for $\theta < t \leq \theta(\mu+\ell)$ by formula 9, section 8.5, p. 255. For some $\theta < 1$, and $\theta(\mu+\ell) \leq t \leq \mu+\ell$, we use formula 4 of section 8.43 on p. 252. In the range $\mu+\ell \leq t \leq \{2(\mu+\ell)\}^2$ see formula 5 of section 8.43. Finally for $t \geq \{2(\mu+\ell)\}^2$ consult equations 1, 2, and 3 on p. 211 of section 7.33 and equation one on p. 74.

LEMMA 6. Let $1/2 < \eta < \mu+\ell+1$. Take ξ, ξ_1, and ξ_3 to be positive numbers. Then

1.9 $$\int_0^{\xi/R} t^{-\eta}|J_{\mu+\ell}(2\pi tR)|\,dt \leq AR^{\eta-1}\,\xi^{\mu+\ell+\eta-1}.$$

Also

1.10 $$\int_{\xi_3/R}^{\infty} t^{-\eta}|J_{\mu+\ell}(2\pi tR)|\,dt \leq A(\ell+1)^6\,\xi_3^{1/2-\eta}\,R^{\eta-1}.$$

If $\xi_1 R < 1$, then

1.11 $$\int_0^{\xi_1} \psi(t)t^{-\eta}|J_{\mu+\ell}(2\pi tR)|\,dt \leq AR^{\mu+\ell}.$$

A does not depend on ξ, ξ_1, or ξ_3. Finally

1.12 $$\int_0^{\infty} t^{-\eta}J_{\mu+\ell}(2\pi tR)\,dt = 2^{-\eta}(2\pi R)^{\eta-1}\Gamma\{\tfrac{1}{2}(\mu+\ell+1-\eta)\}\{\Gamma\{\tfrac{1}{2}(\mu+\ell+1+\eta)\}\}^{-1}.$$

Proof: 1.9 and 1.11 are consequences of 1.7. 1.10 follows from 1.8. For 1.12, see [4], p. 39, Lemma 2.6.3.

LEMMA 7. Assume $b(t)$ is slowly varying. Let $1/2 < \eta < \mu+\ell+1$. Then

1.13
$$\int_0^\infty \psi(t)t^{-\eta}b(t)J_{\mu+\ell}(2\pi tR)dt$$
$$= \tfrac{1}{2}\pi^{\eta-1}\Gamma\{\tfrac{1}{2}(1+\mu+\ell-\eta)\}\{\Gamma\{\tfrac{1}{2}(1+\mu+\ell+\eta)\}\}^{-1}R^{\eta-1}b(1/R) + E_0(R).$$

For any $\sigma > 0$, there is an R_0 not depending on ℓ, such that

1.14
$$|E_0(R)| < \sigma(\ell+1)^6 R^{\eta-1}b(1/R) + AR^{\ell+\mu}$$

for $R < R_0$.

(The proof of Lemma 7 follows the lines of the proof of Theorem 6.2, p. 187 [25], Vol. I.)

Proof: Let σ_1 be an arbitrary positive number. Let $\delta > 0$ be so small that $1/2 < \eta-\delta$ and so small that $\eta+\delta < \mu+\ell+1$. Using Lemma 1 and the definition of slowly varying function, we see there is a $\xi_1 > 1$ such that for $t \geq \xi_1$ the following three conditions hold:

i) $b(t)$ is of constant sign.
ii) $t^\delta|b(t)|$ is increasing.
iii) $t^{-\delta}|b(t)|$ is decreasing.

We may assume without loss of generality $b(t) \geq 0$ for $t \geq \xi_1$. Then for $t \geq \xi_1$, $t^\delta b(t)$ is increasing, and $t^{-\delta}b(t)$ is decreasing.

We now choose two numbers ξ_2 and ξ_3 with $0 < \xi_2 \leq 1 \leq \xi_3$. ξ_2 and ξ_3 will not depend on R or ℓ. We assume $R < \xi_2/\xi_1$, and $R < \xi_1^{-1}$. Then $t^{-\delta}b(t)$ is decreasing for $t \geq \xi_2/R$. Since $\xi_3 \geq \xi_2$, we also know that $t^{-\delta}b(t)$ is decreasing for $t \geq \xi_3/R$. For the remainder of the proof of Lemma 7, we shall let J denote $J_{\mu+\ell}(2\pi tR)$. Then

1.15
$$\int_0^\infty \psi(t)t^{-\eta}b(t)J\,dt = b(1/R)\int_0^\infty t^{-\eta}J\,dt + E_0(R).$$

1.16 $$E_0(R) = -b(1/R)\int_0^{\xi_2/R} J t^{-\eta}dt - b(1/R)\int_{\xi_3/R}^{\infty} J t^{-\eta}dt$$

$$+\int_{\xi_1}^{\xi_2/R} b(t)Jt^{-\eta}dt + \int_0^{\xi_1} \psi(t)b(t)Jt^{-\eta}dt + \int_{\xi_3/R}^{\infty} b(t)Jt^{-\eta}dt$$

$$+\int_{\xi_2/R}^{\xi_3/R} \{b(t) - b(1/R)\}Jt^{-\eta}dt.$$

[We have used the fact that $\psi(t) = 1$ for $t \geq \xi_1$].

We now proceed to estimate the various terms of E_0.

$$\left|\int_{\xi_1}^{\xi_2/R} t^{\delta} b(t)J t^{-\eta-\delta}dt\right| \leq R^{-\delta} b(1/R)\int_0^{\xi_2/R} |Jt^{-\eta-\delta}|\, dt,$$

since $t^{\delta}b(t)$ is increasing for $t \geq \xi_1$. Hence by equation 1.9 of Lemma 6, (with the η of 1.9 here taken to be $\eta+\delta$ and with $\xi = \xi_2$) we have

1.17 $$\left|\int_{\xi_1}^{\xi_2/R} b(t)J t^{-\eta}dt\right| \leq Ab(1/R)R^{\eta-1}\xi_2^{\ell+\mu+1-\eta-\delta}.$$

Also

$$\left|\int_{\xi_3/R}^{\infty} b(t)Jt^{-\eta}dt\right| = \left|\int_{\xi_3/R}^{\infty} t^{-\delta}b(t)Jt^{-\eta+\delta}dt\right| \leq R^{\delta}b(1/R)\int_{\xi_3/R}^{\infty} |J|\, t^{-\eta+\delta}dt$$

since $t^{-\delta}b(t)$ is decreasing for $t \geq \xi_3/R$. So by equation 1.10 of Lemma 6

1.18 $$\left|\int_{\xi_3/R}^{\infty} b(t)Jt^{-\eta}dt\right| \leq A(\ell+1)^6\, b(1/R)R^{\eta-1}\xi_3^{\eta-1/2+\delta}.$$

(We have taken the η of 1.10 to be $\eta-\delta$ here). Next,

$$\left|\int_{\xi_2/R}^{\xi_3/R} \{b(t)-b(1/R)\}Jt^{-\eta}dt\right| \leq \max_{\xi_2/R\leq t\leq\xi_3/R} |b(t)-b(1/R)|\int_0^{\xi_3/R} |Jt^{-\eta}|\,dt$$

$$\leq A \max_{\xi_2/R\leq t\leq\xi_3/R} |b(t) - b(1/R)|\, R^{\eta-1}\xi_3^{\ell+\mu+1-\eta}$$

by equation 1.9 in Lemma 6. (We have taken ξ in equation 1.9 to be ξ_3.) Now by Lemma 2, we have

1.19 $$\left|\int_{\xi_2/R}^{\xi_3/R} \{b(t) - b(1/R)\} J t^{-\eta} dt\right| \leq \sigma_1 b(1/R) R^{\eta-1} \xi_3^{\mu+\ell+1-\eta}$$

for $R \leq R_1(\sigma_1, \xi_2, \xi_3)$. Using once again equations 1.9 and 1.10 respectively, we find

1.20 $$\left|b(1/R)\int_0^{\xi_2/R} J t^{-\eta} dt\right| \leq AR^{\eta-1} b(1/R) \xi_2^{\ell+\mu+1-\eta},$$

and

1.21 $$\left|b(1/R)\int_{\xi_3/R}^{\infty} J t^{-\eta} dt\right| \leq A(\ell+1)^6 b(1/R) R^{\eta-1} \xi_3^{1/2-\eta}.$$

Finally by 1.10 of Lemma 6, we see

1.22 $$\left|\int_0^{\xi_1} \psi(t) b(t) J t^{-\eta} dt\right| \leq AR^{\mu+\ell}.$$

We now substitute from equations 1.17, 1.18, 1.19, 1.20, 1.21 and 1.22 into equation 1.16.

We obtain, for any $\sigma_1 > 0$,

$$|E_0(R)| \leq AR^{\eta-1} b(1/R) \xi_2^{\ell+\delta+1-\eta-\delta}$$

$$+ A(\ell+1)^6 R^{\eta-1} b(1/R) \xi_3^{1/2-\eta+\delta} + AR^{\ell+\mu} + \sigma_1 R^{\eta-1} b(1/R) \xi_3^{\mu+\ell+1-\eta},$$

for $R < R_0(\sigma_1, \xi_2, \xi_3)$. A does not depend on R, ℓ, ξ_1, ξ_2 or ξ_3.

We now obtain 1.14 by first choosing ξ_2 sufficiently small and ξ_3 sufficiently large; then we choose σ_1 sufficiently small. The proof of Lemma 7 now follows from 1.15 and equation 1.12 of Lemma 6.

Proof of Theorem 1: During the proof of Theorem 1, we shall write J for $J_{\frac{1}{2}(k-2)+\ell+\nu}(2\pi Rt)$ for simplification in printing. This notation begins in equation 1.26. (ν will be defined later.)

By Lemma 3,

1.24 $$F_\varepsilon(x) = 2\pi(-i)^\ell R^{\frac{1}{2}\{2-k\}} Y_\ell(x') \int_0^\infty \psi(t) t^{-\gamma+k/2} b(t) \cdot \exp(-\varepsilon t) J_{\frac{1}{2}(k-2)+\ell}(2\pi R t)\, dt.$$

We wish to show first of all that $F(x) = \lim_{\varepsilon \to 0+} F_\varepsilon(x)$ exists for $x \neq 0$, and $D^s F(x)$ exists for any $s = (s_1, \ldots, s_k)$, (s_j integers). We shall apply Lemma 4 to the integral in 1.24. We shall take $\nu = \max\{1 + \|s\|, [\frac{1}{2}k + 3]\}$ if $\nu \geq \frac{1}{2}k$. Otherwise set $\nu = \max\{[\frac{1}{2}k - \gamma + 2 + \|s\|], [\frac{1}{2}k + 3]\}$. Here $\|s\| = s_1 + s_2 + \ldots + s_k$, and $[\alpha]$ denotes the largest integer equal to or less than α. Also we shall set the η of Lemma 4 equal to $\gamma - \frac{1}{2}k$, and we put $\mu = \frac{1}{2}(k-2)$. (We remark that we shall not use the fact that $\nu \geq \frac{1}{2}k + 2$ until we investigate $F^\wedge(x)$ for those $F(x)$ in L^1.) Applying Lemma 4 to the integral in 1.24, we find

1.25 $$F_\varepsilon(x) = 2\pi(-i)^\ell R^{\frac{1}{2}(2-k)} Y_\ell(x') K_\varepsilon(R) + E_{\varepsilon,1}(x).$$

1.26 $$K_\varepsilon(R) = (\pi R)^{-\nu} (\tfrac{1}{2}(\gamma+\ell))_\nu \int_0^\infty \psi(t) t^{-\nu-\gamma+\frac{1}{2}k} b(t) \exp(-\varepsilon t) J\, dt + (\pi R)^{-\nu} \sum_{\substack{m+p+q=\nu \\ q<\nu}} d(m, p, o, q)\, \varepsilon^m \int_0^\infty \psi(t) t^{-\eta-p-q} h_p(t) \exp(-\varepsilon t) J\, dt;$$

and

1.27 $$E_{\varepsilon,1}(x) = 2\pi^{1-\nu}(-i)^\ell Y_\ell(x') R^{\frac{1}{2}(2-k)-\nu} \sum_{\substack{m+p+g+q=\nu \\ g>0}} d(m, p, g, q)\, \varepsilon^m \int_0^\infty \psi^{(g)}(t)\, t^{-\eta-p-q} h_p(t) \exp(-\varepsilon t)\, J\, dt.$$

We first look at $E_{\varepsilon,1}$. Since $\psi^{(g)}(t) = 0$ for $t \geq 1$, it is clear that $E_1(x) = \lim_{\varepsilon \to 0+} E_{\varepsilon,1}(x)$ exists. Furthermore,

1.28 $$E_1(x) = 2\pi^{1-\nu}(-i)^{\ell} Y_{\ell}(x')R^{\frac{1}{2}(2-k)-\nu}$$

$$\cdot \sum_{\substack{p+g+q=\nu \\ g>0}} d(o,p,g,q) \int_0^1 \psi^{(g)}(t)t^{-\eta-p-q} h_p(t)J\,dt.$$

$\psi^{(g)}(t) = 0$ for $t \leq \frac{1}{2}$, and $J_{\frac{1}{2}(k-2)+\ell+\nu}(2\pi Rt)$ is infinitely differentiable in R for $R > 0$ and $t \geq 1/2$. Therefore it is clear from equation 1.28 that $E_1(x)$ is infinitely differentiable for $x \neq 0$.

Now in order to show $\lim_{\varepsilon \to 0+} F_\varepsilon(x)$ exists for $x \neq 0$, we must prove $\lim_{\varepsilon \to 0+} K_\varepsilon(R)$ exists for $R \neq 0$. Hence we need to demonstrate

$$\lim_{\varepsilon \to 0+} \varepsilon^m \int_0^\infty \psi(t)t^{-\eta-p-q} h_p(t) \exp(-\varepsilon t)J\,dt$$

exists for $R \neq 0$. (We include the case $p = 0$, i.e., $h_0(t) = b(t)$).

Here $\nu = m+p+q$, and $\nu \geq \frac{1}{2}k - \gamma + 1 = -\eta + 1$. Since $h_p(t)$ is slowly varying, $|h_p(t)| \leq A(t^{1/4}+1)$. Also by equation 1.8 of Lemma 5, $|J| \leq A(\ell+1)^6(tR)^{-1/2}$. Furthermore $\varepsilon^m \exp(-\varepsilon t) = O(t^{-m})$. Thus

$$|\varepsilon^m \psi(t)t^{-\eta-p-q} h_p(t)\exp(-\varepsilon t)J| = O(1+t)^{-5/4}$$

for fixed $R > 0$. Hence the above limit exists as $\varepsilon \to 0+$ by dominated convergence. Moreover those terms for which $m \geq 1$ tend to zero as $\varepsilon \to 0+$. To see this use the fact that for $m \geq 1$, $\varepsilon^m \exp(-\varepsilon t) \leq A\varepsilon^{1/8} t^{-m+1/8}$ instead of $\varepsilon^m \exp(-\varepsilon t) \leq At^{-m}$.

Therefore, we see that $F(x) = \lim_{\varepsilon \to 0+} F_\varepsilon(x)$ exists for $x \neq 0$, and

1.29 $$F(x) = 2\pi^{1-\nu}(-i)^{\ell} R^{\frac{1}{2}(2-k)-\nu}(\tfrac{1}{2}(\gamma+\ell))_\nu \cdot \int_0^\infty \psi(t)t^{-\nu-\gamma+\frac{1}{2}k} b(t)J\,dt$$

$$+ 2\pi^{1-\nu}(-i)^{\ell} Y_{\ell}(x')R^{\frac{1}{2}(2-k)-\nu} \sum_{p+q=\nu} d(o,o,p,q) \cdot \int_0^\infty \psi(t)t^{-\nu-\gamma+\frac{1}{2}k} h_p(t)J\,dt + E_1(x).$$

$E_1(x)$ is infinitely differentiable for $x \neq 0$. Also from the definition of spherical harmonic, it is clear that $Y_n(x')$ is infinitely differentiable for $x \neq 0$. (See [6], p. 31, top of page.) Therefore we can prove $D^s F$ exists for $x \neq 0$ by showing

1.30 $$d^{\|s\|}/dR^{\|s\|}\left\{\int_0^\infty \psi(t)t^{-\nu-\gamma+\frac{1}{2}k}\, h_p(t)\, J\, dt\right\}$$

exists for $R \neq 0$, $p = 0, 1, \ldots, \nu$. Now

1.31 $$d/dR\,\{J_B(2\pi tR)\} = \pi t\{J_{B-1}(2\pi Rt) - J_{B+1}(2\pi Rt)\}.$$

(See [24], p. 45, equation 2.) Therefore, for any $j \geq 0$,

1.32 $$\frac{d^j}{dR^j} J = (\pi t)^j \sum_{c=-j}^{j} e_c\, J_{\frac{1}{2}(k-2)+\ell+\nu+c}(2\pi tR).$$

The e_c do not depend on ℓ. Also, if

$j \leq \|s\|$, $\frac{1}{2}(k-2) + \ell + \nu + c \geq \frac{1}{2}(k-2) + \ell \geq -\frac{1}{2}$. Finally $\nu > 1 + \|s\| + \frac{1}{2}k - \gamma$.

Therefore, by Lemma 6,

$$\int_0^\infty \psi(t)t^{-\nu-\gamma+\frac{1}{2}k}(1+t)^{\|s\|}\, |h_p(t)| \cdot |J_{\frac{1}{2}(k-2)+\ell+\nu+c}(2\pi Rt)|\, dt < \infty.$$

Hence we can easily justify $\|s\|$ differentiations under the integral sign in 1.30, and we conclude that $D^s F(x)$ exists for $x \neq 0$. As a matter of fact, our argument shows more. Namely if x lies in a compact set Ω which does not contain the origin, then

1.33 $$\left|d^{\|s\|}/dR^{\|s\|}\int_0^\infty \psi(t)t^{-\nu-\gamma+\frac{1}{2}k} h_p(t) J\, dt\right| \leq A(\Omega, \|s\|)(\ell+1)^6.$$

For the moment we shall bypass the proof of ii), and we shall turn to the behavior of $F(x)$ near $x = 0$. Since $b(t)$ is in S, all of the $h_p(t)$ are slowly varying. Hence by Lemma 7

1.34 $$\int_0^\infty \psi(t)t^{-\nu-\gamma+\frac{1}{2}k} h_p(t) J\, dt$$

$$= \tfrac{1}{2}\pi^{\nu+\gamma-\frac{1}{2}k-1}\Gamma\{\tfrac{1}{2}(\ell+2k-\gamma)\}\{\Gamma\{\tfrac{1}{2}(2\nu+\ell+\gamma)\}\}^{-1}$$

$$\cdot\; R^{\nu+\gamma-1-\frac{1}{2}k} h_p(1/R) + E_0(r).$$

For any $\sigma > 0$, there exists an R_0 not depending on ℓ such that

1.35 $$|E_0(R)| < \sigma R^{\gamma+\nu-\frac{1}{2}(k+2)} b(1/R)(\ell+1)^6 + AR^{\frac{1}{2}(k-2)+\ell+\nu}$$

for $R \leq R_0$.

It is clear from 1.28 and 1.7 of Lemma 5 that

1.36 $$|E_1(x)| \leq A \max_{x' \text{ in } S_{k-1}} |Y_\ell(x')| \max |d(m, g, p, q)| R^\ell.$$

For $p \geq 1$, Lemma 1 implies $|h_p(1/R)| = o(b(1/R))$ as $R \to 0$. Therefore equations 1.1 and 1.2 follow from 1.29, 1.35 and 1.36. Moreover, if we use the fact that $|d(m, g, p, q)| \leq A(1+\ell)^\nu$ (see Lemma 4), we obtain the following estimate on $E(x)$.

1.37 $$|E(x)| = o(1) \max_{x' \text{ in } S_{k-1}} |Y_\ell(x')| (1+\ell)^{\nu+6} \{R^{\gamma-k} b(1/R) + AR^\ell\}.$$

The little o does not depend on ℓ.

We now turn to statement ii) of Theorem 1. First of all, we assert that if r is a sufficiently large positive integer,

1.38 $$\|\Delta^r\{\psi(|x|)|x|^{-\gamma} Y_\ell(x')b(|x|)\exp(-\varepsilon|x|)\}\|_1 \leq M < \infty,$$

with M independent of ε. (Δ is the Laplacean operator.) There is no difficulty differentiating at the origin because $\psi(|x|) = 0$ for $|x| \leq \frac{1}{2}$, and this is the whole point of introducing the function ψ. Now for any $s = (s_1, s_2, \ldots, s_k)$,

$$|D^s\{|x|^\ell Y_\ell(x')\}| \leq A(s)|x|^{\ell-\|s\|}(1+\ell)^{\frac{1}{2}(k-1)+\|s\|}\cdot\|Y_\ell(x')\|_2.$$

See [6], p. 33, Theorem 4. We also recall that $b'(t) = t^{-1}h_1(t)$, $h_1'(t) = t^{-1}h_2(t)\ldots$ with the $h_j(t)$ slowly varying. Thus, using the fact that $\varepsilon^m \exp(-\varepsilon t) \leq A(m)t^{-m}$, one easily sees that

1.39 $$\|\Delta^r\{\psi(|x|)|x|^{-\gamma}Y_\ell(x')b(|x|)\exp(-\varepsilon|x|)\}\|_1 \leq A(\gamma, r)(1+\ell)^{2r+\frac{1}{2}(k-1)}\|Y_\ell(x')\|_2$$

if $r > \frac{1}{2}(k+1-\gamma)$. (The L^2 norm of Y_ℓ is taken over S_{k-1}).

Now let $\theta(x)$ be in $C^\infty(E_k)$. In addition, assume $\theta(x)$ and all of its derivatives are in L^1. Then for any integer r

$$(-|x|)^{2r}\theta^\vee(x) = (2\pi)^{-2r}\{\Delta^r\theta\}^\vee.$$

See (4), p. 45, equation 2.8.9. Therefore,

1.40 $$|\theta^\vee(x)| \leq \{2\pi|x|\}^{-2r}\|\Delta^r\theta\|_1.$$

We take

$$\theta(x) = \psi(|x|)|x|^{-\gamma}Y_\ell(x')b(|x|)\exp(-\varepsilon|x|).$$

Applying 1.39 we find

1.41 $$|F_\varepsilon(x)| \leq A(\gamma, r)(1+\ell)^{2r+\frac{1}{2}(k-1)}|x|^{-r}\|Y_\ell(x')\|_2$$

if $r > \frac{1}{2}(k+1-\gamma)$. This finishes the proof of ii).

We shall now assume $F(x)$ is in L^1, and we shall show

$$F^\wedge(x) = \psi(|x|)|x|^{-\gamma}b(|x|)Y_\ell(x').$$

We have already proven 1.1 and 1.2. It follows easily from 1.1 and 1.2 and Lemma 1(i) that if $F(x)$ is in L^1, then

1.42 $$\int_0^\infty \psi(t)t^{-\gamma-1}b(t)dt < \infty.$$

Our main step will be to use 1.42 to find a function $H(x)$ in L^1 such that $|F_\varepsilon(x)| \leq H(x)$. We shall first find a majorant for $K_\varepsilon(R)$. ($K_\varepsilon(R)$ is defined in

1.26.) Let

$$B = \max_{\substack{\xi, j \\ j=0,1,\ldots \nu \\ 0 \le \xi < \infty}} \xi^j e^{-\xi}.$$

Define

1.43 $$K(R) = (\tfrac{1}{2}(n+\gamma))_\nu (\pi R)^{-\nu} \int_0^\infty \psi(t) t^{-\nu-\gamma+\frac{1}{2}k} |b(t)|\,|J|\,dt$$

$$+ B(\pi R)^{-\nu} \sum_{\substack{m+p+q=\nu \\ q<\nu}} |d(m,p,o,q)| \int_0^\infty \psi(t) t^{-\nu-\gamma+\frac{1}{2}k} |h_p(t)|\,|J|\,dt.$$

Since $\varepsilon^m \exp(-\varepsilon t) \le Bt^{-m}$, equation 1.26 implies $|K_\varepsilon(R)| \le K(R)$. Therefore

1.44 $$|2\pi(-i)^\ell R^{\frac{1}{2}(2-k)} Y_\ell(x') K_\varepsilon(R)| \le AC(\ell) R^{\frac{1}{2}(2-k)} K(R),$$

where $C(\ell) = \max_{x' \text{ in } S_{k-1}} |Y_\ell(x')|$. Let

$$E_3(x) = C(\ell) R^{-\nu+\frac{1}{2}(2-k)} \cdot \sum_{\substack{m+p+g+q=\nu \\ q<\nu \\ g>0}} |d(m,p,g,q)|$$

$$\cdot \int_0^1 |\psi^{(g)}(t)|\, t^{-\gamma+\frac{1}{2}k-p-q} |h_p(t)|\,|J|\,dt.$$

A glance at equation 1.27 shows

1.45 $$|E_{\varepsilon,1}(x)| \le E_3(x).$$

Now by 1.25, 1.44, and 1.45, we see

1.46 $$|F_\varepsilon(x)| \le H(x)$$

where

$$H(x) = AC(\ell) R^{\frac{1}{2}(2-k)} K(R) + E_3(x).$$

We shall now verify that $H(x)$ is in L^1. It is clear that $E_3(x)$ is in L^1

since $\nu > 1 + \frac{1}{2}(k-2)$. Also

$$\int_{E_k} K(|x|)|x|^{1-\frac{1}{2}k}dx \leq A\int_0^\infty R^{\frac{1}{2}k} K(R)dR.$$

By Lemma 1, $|h_p(t)| \leq |b(t)|$ for large t, so by looking at 1.43 we see that it suffices to show

1.47 $$\int_0^\infty R^{\frac{1}{2}k-\nu} \int_0^\infty \psi(t)t^{-\nu-\gamma+\frac{1}{2}k}|b(t)|\,|J|\,dtdR < \infty .$$

By Fubini's theorem the left hand side of 1.47 is

$$\int_0^\infty \psi(t)t^{-\nu-\gamma+\frac{1}{2}k}|b(t)| \int_0^\infty R^{\frac{1}{2}k-\nu}\,|J|\,dt\,dR$$

$$= \int_0^\infty \psi(t)|b(t)|t^{-1-\gamma}dt \int_0^\infty u^{\frac{1}{2}k-\nu}\,|J_{\frac{1}{2}k+\ell+\nu-1}(2\pi u)|\,du.$$

In the last step we made a change of variables, $u = Rt$. By 1.42

$$\int_0^\infty \psi(t)|b(t)|t^{-1-\gamma}dt < \infty.$$

Since $\nu \geq 2 + \frac{1}{2}k$, we see using Lemma 6 that

$$\int_0^\infty u^{\frac{1}{2}k-\nu}\,|J_{\frac{1}{2}k+\ell+\nu-1}(2\pi u)|\,du < \infty.$$

Thus $H(x)$ is in L^1.

Now by the Fourier inversion formula (see [4], p. 25, equation 2.1.19)

$$\int_{E_k} \exp(-2\pi i x\cdot y)F_\varepsilon(y)dy = \psi(|x|)|x|^{-\gamma}b(|x|)\exp(-\varepsilon|x|).$$

Hence by dominated convergence

$$F^\wedge(x) = b(|x|)\psi(|x|)|x|^{-\gamma}.$$

This concludes the proof of Theorem 1 proper, and we turn to a discussion of the remarks following Theorem 1.

Remark 1 is a trivial consequence of equations 1.1, 1.2 and assertion ii) of Theorem 1.

Proof of Remark 2. If $b(t) \equiv 1$

$$F_\varepsilon(x) = \{|x|^{-\gamma}\exp(-\varepsilon|x|)Y_\ell(x')\}^\vee + \{|x|^{-\gamma}[\psi(|x|) - 1]\exp(-\varepsilon|x|)Y_\ell(x')\}^\vee.$$

Since $|x|^{-\gamma}[\psi(|x|) - 1]Y_\ell(x')$ has compact support

$$E(x) = \lim_{\varepsilon \to 0+} \{|x|^{-\gamma}[\psi(|x|) - 1]\exp(-\varepsilon|x|)Y_\ell(x')\}^\vee$$

exists and is infinitely differentiable. Furthermore, any derivative of $E(x)$ is bounded.

Also, by Lemma 3

$$\{|x|^{-\gamma}\exp(-\varepsilon|x|)Y_\ell(x')\}^\vee = (2\pi)(-i)^\ell R^{\frac{1}{2}(2-k)}$$

$$\cdot\, Y_\ell(x') \cdot \int_0^\infty t^{\frac{1}{2}k-\gamma}\exp(-\varepsilon t)J_{\frac{1}{2}(k-2)+\ell}(2\pi Rt)dt.$$

Remark 2 now follows because

$$\lim_{\varepsilon \to 0+} \int_0^\infty t^{\frac{1}{2}k-\gamma}\exp(-\varepsilon t)J_{\frac{1}{2}(k-2)+\ell}(2\pi Rt)dt$$

$$= \tfrac{1}{2}(\pi R)^{\gamma-\frac{1}{2}(k+2)}\Gamma\{\tfrac{1}{2}(k+\ell-\gamma)\}\{\Gamma\{\tfrac{1}{2}(\ell+\gamma)\}\}^{-1}.$$

See [4], p. 39, Lemma 2.6.3.

We now turn to remark 3. We shall need two facts about spherical harmonics: (See [6], p. 33, Theorem 4.) If

$$\int_{S_{k-1}} |Y_\ell(x')|^2 d\sigma = 1$$

then

1.48 $$|Y_\ell(x')| \leq A\ell^{\frac{1}{2}k-1},$$

and

1.49 $$|D^s Y_\ell(x')| \le A(\Omega, s)\ell^{\frac{1}{2}k-1+\|s\|}$$

for x in Ω. (Ω is any compact set of E_k not containing the origin.) Assertion i) follows from 1.37 and 1.48. Statement iii) is a consequence of 1.41. iv) follows from 1.1, i) and iii). ii) can be seen from 1.49, 1.28, 1.29, and 1.33 if one keeps in mind the fact that $|d(m, p, g, q)| \le A(1+\ell)^\nu$.

We now need to recall certain facts about spherical harmonics. (See chapter 6 of [6]). Any $W(x')$ in $C^\infty(S_{k-1})$ has an expansion in spherical harmonics

1.50 $$W(x') = \sum_{\ell, m} a(\ell, m) Y_{\ell, m}(x').$$

$Y_{\ell, m}(x')$ is a spherical harmonic of degree ℓ, and

$$\int_{S_{k-1}} |Y_{\ell, m}(x')|^2 \, d\sigma = 1.$$

ℓ runs through the non-negative integers, and for each ℓ there are less than $A_k(1+\ell)^{k-2}$ values of m. Moreover, for infinitely differentiable $W(x')$, and any integer j,

1.51 $$|a_{\ell, m}| \le A_j(1+\ell)^{-j}.$$

To each $W(x')$ in $C^\infty(S_k)$, we associate functions $\tilde{W}_{\gamma, k}(x')$. If $W(x')$ is defined by 1.50

1.52 $$\tilde{W}_{\gamma, k}(x') = \pi^{\gamma - k} \sum_{\ell, m} (-i)^\ell a(\ell, m) \cdot \Gamma\{\tfrac{1}{2}(k+\ell-\gamma)\}\{\Gamma\{\tfrac{1}{2}(\ell+\gamma)\}\}^{-1} Y_{\ell, m}(x').$$

$\tilde{W}_{\gamma, k}(x')$ is also in $C^\infty(S_{k-1})$. (One can formally differentiate under the sum sign in 1.52 because of 1.49 and 1.51.) (If $\gamma \ge \frac{1}{2}k$, the effect of the factor $\Gamma\{\frac{1}{2}(k+\ell-\gamma)\}\{\Gamma\{\frac{1}{2}(\ell+\gamma)\}\}^{-1}$ is closely allied to spherical fractional integration of order $\gamma - \frac{1}{2}k$. See [11]. If $\gamma < \frac{1}{2}k$, let ℓ be any integer larger than $\frac{1}{2}(\frac{1}{2}k-\gamma)$. The effect of the factor $\Gamma\{\frac{1}{2}(k+\ell-\gamma)\}\{\Gamma\{\frac{1}{2}(\ell+\gamma)\}\}^{-1}$ is essentially the effect of using the "spherical Laplacean" ℓ times followed by fractional spherical integration of order $2\ell + \gamma - \frac{1}{2}k$. The spherical Laplacean is the operator L defined on the top of p. 34 in [6].)

Theorem 2 will be a generalization of Theorem 1 in that $Y_\ell(x')$ will be replaced by any $W(x')$ in $C^\infty(S_{k-1})$.

THEOREM 2. Let $W(x')$ be in $C^\infty(S_{k-1})$, and suppose the expansion of W in spherical harmonics is given by 1.50. I.e.,

$$W(x') \sim \sum a(\ell, m) Y_{\ell, m}(x')$$

Assume $\gamma + \ell \neq 0, -2, -4, \ldots$ and $\gamma < k + \ell$ for all non-negative integers ℓ for which $a(\ell, m) \neq 0$. Let b and ψ be the same as in Theorem 1. Define

$$F_\varepsilon(x) = \{\psi(|y|)|y|^{-\gamma} b(|y|)W(y') \exp(-\varepsilon |y|)\}^{\vee}.$$

Then conclusions i) and ii) of Theorem 1 hold. At $x = 0$, we have

$$F(x) = \tilde{W}_{\gamma, k}(x') b(1/R) R^{\gamma - k} + o\{b(1/R)\} + O(R^{\Lambda}).$$

Λ is the smallest integer such that $\ell < \Lambda$ implies $a(\ell, m) = 0$ for all m. $\tilde{W}_{\gamma, k}$ is defined by 1.52.

Also if $F(x)$ is in L^1,

$$F^{\wedge}(x) = \psi(|x|)|x|^{-\gamma} b(|x|) W_{\gamma, k}(x').$$

<u>Proof:</u> To prove Theorem 2, we expand $W(x')$ into $\sum a(\ell, m) Y_{\ell, m}(x')$. We then write

$$F_\varepsilon(x) = \sum_{\ell, m} F_{\varepsilon, \ell, m}(x)$$

with

$$F_{\varepsilon, \ell, m}(x) = \{a(\ell, m)\psi(|y|)|y|^{-\gamma} b(|y|) Y_{\ell, m}(y') \exp(-\varepsilon |y|)\}^{\vee}.$$

In view of Remark 3 to Theorem 1, and in view of our discussion of spherical harmonics (especially 1.51), we can deduce Theorem 2 by applying Theorem 1 to each $F_{\varepsilon, \ell, m}$.

In Theorems 3 and 4, we treat the exceptional cases that $\ell + \gamma$ is a non-positive even integer and $\gamma = k + \ell$. The most important case of Theorem 3 is the case $\gamma = \ell = 0$.

THEOREM 3. Let $\gamma + \ell = -2j$ for some non-negative integer j. Define $F_\varepsilon(x)$ as in Theorem 1. Then conclusions i) and ii) of Theorem 1 hold.
At $x = 0$,

1.53 $$F(x) = AR^{\gamma-k-1} b'(1/R) Y_\ell(x') + o\{R^{\gamma-k-1} b'(1/R)\} + O(1).$$

$A \neq 0$.

Now suppose $\ell = \gamma = 0$. Then if $b(t) \to 0$ as $t \to \infty$, $F(x)$ is in L^1, and

$$F\hat{}(x) = \psi(|x|)b(|x|).$$

<u>Proof:</u> To prove Theorem 3, we need a lemma analogous to Lemma 4.

LEMMA 8. Let ν be an integer equal to or greater than one. Assume $b(t)$ is in S. Let $\eta = -1 - \mu - \ell - 2\sigma$ for some integer σ with $0 \leq \sigma \leq \nu-1$. Then

1.54 $$\int_0^\infty \psi(t)t^{-\eta}b(t)\exp(-\varepsilon t) J_{\mu+\ell}(2\pi tR)dt$$

$$= A_1 R^{-\nu} \int_0^\infty \psi(t)t^{-\eta-\nu+1} b'(t)\exp(-\varepsilon t) J_{\mu+\ell+\nu}(2\pi tR)dt + R^{-\nu}\sum e(m, p, g, q)\varepsilon^m$$

$$\cdot \int_0^\infty \psi(t)t^{-\eta-p-q} h_p(t) \exp(-\varepsilon t) J_{\mu+\ell+\nu}(2\pi tR)\, dt.$$

$A_1 \neq 0$. The sum in 1.53 is extended over non-negative integers m, g, p, and q such that $m+g+p+q = \nu$ except that if q is ν or $\nu-1$, $m \geq 1$ or $g \geq 1$ (or both).

REMARK. The important difference between Lemma 8 and Lemma 4 is that in Lemma 8 no term occurs with $p = \nu$, $m = 0$, and $g = 0$.

<u>Proof:</u> We apply Lemma 4 to the left side of 1.54; we take the ν of Lemma 4 to be our present σ. We then have,

1.55 $$\int_0^\infty \psi(t)t^{-\eta}\, b(t)\exp(-\varepsilon t)J_{\mu+\ell}(2\pi tR)dt$$

$$= (\pi R)^{-\sigma}(\tfrac{1}{2}(\eta+\ell+\mu+1))_\sigma \int_0^\infty \psi(t)t^{-\sigma-\eta}\exp(-\varepsilon t)\,J_{\mu+\sigma+\ell}(2\pi tR)\,dt$$

$$+ (\pi R)^{-\sigma} d(0,1,0,\sigma-1)\int_0^\infty \psi(t)t^{-\eta-\sigma}h_1(t)\exp(-\varepsilon t)\,J_{\mu+\ell+\sigma}(2\pi tR)\,dt$$

$$+ (\pi R)^{-\sigma}\sum d(m,p,g,q)\,\varepsilon^m \int_0^\infty \psi(t)t^{-\eta-p-q}h_p(t)\exp(-\varepsilon t)\,J_{\mu+\ell+\sigma}(2\pi tR)\,dt.$$

The sum is extended over non-negative integers m, p, g, and q such that $m+p+g+q=\sigma$ except that $g \geq 1$ or $m \geq 1$ if $p = 0$ or 1. We must now consider what happens to the first two terms on the right hand side of 1.55 when we integrate by parts once more. Since $\eta = \mu - \ell - 2\sigma - 1$,

$$t^{-\eta-\sigma}J_{\mu+\ell+\sigma}(2\pi tR) = t^{\mu+\ell+\sigma+1}\,J_{\mu+\ell+\sigma}(2\pi tR)$$

$$= (2\pi R)^{-\mu-\ell-\sigma-2}d/dt\{(2\pi R)^{\mu+\ell+\sigma+1}\,J_{\mu+\ell+\sigma+1}(2\pi tR)\}.$$

(We have used again the recurrence relations for Bessel functions. See [24], p. 45, sixteenth line from top of page.)

Therefore integrating by parts on the right-hand side of 1.55, we find

1.56 $$\int_0^\infty \psi(t)t^{-\eta}b(t)\exp(-\varepsilon t)\,J_{\mu+\ell}(2\pi tR)dt$$

$$= \tfrac{1}{2}(\pi R)^{-\sigma-1}(\tfrac{1}{2}(\eta+\mu+\ell+1))_\sigma \int_0^\infty \psi(t)t^{-\eta-\sigma-1}\,h_1(t)\exp(-\varepsilon t)J_{\mu+\ell+\sigma+1}(2\pi tR)dt$$

$$+ \tfrac{1}{2}(\pi R)^{-\sigma-1}d(0,1,0,\sigma-1)\int_0^\infty \psi(t)t^{-\eta-\sigma-1}h_2(t)\cdot\exp(-\varepsilon t)J_{\mu+\ell+\sigma+1}(2\pi tR)dt$$

$$+ (\pi R)^{-\sigma-1}\sum d_1(m,p,g,q)\,\varepsilon^m\int_0^\infty \psi^{(g)}(t)t^{-\eta-p-q}h_p(t)\exp(-\varepsilon t)J_{\mu+\ell+\sigma+1}(2\pi tR)dt.$$

The above sum is extended over non-negative values of m, p, g, and q, such that $m+p+g+q=\sigma+1$ except there are no terms with $p = 1$ or 2, and $g = m = 0$.

The $d_1(m, p, g, q)$ are some numbers. We now finish the proof by $\nu - \sigma - 1$ more stages of integration by parts as in the proof of Lemma 4.

Proof of Theorem 3: If we use Lemma 8 in place of Lemma 4, we prove assertions i) and ii), and equation 1.53 in exactly the same way we proved statements i), ii), and equation 1.1 in Theorem 1.

We now assume $\gamma = \ell = 0$, and $b(t)$ tends to zero as t tends to infinity. We wish to show $F(x)$ is in L^1, and

$$F\hat{}(x) = \psi(|x|)\, b(|x|).$$

One could base the proof of these facts on the following lemma due to Rudin [19]:

> "Choose an integer $p > \frac{1}{2}(n+1)$. Suppose g is continuous on $0 \leq t < \infty$, $g(t) \to 0$ as $t \to \infty$, and $(-1)^k \delta^k g(t) > 0$, $(k = 0, 1, \ldots, p;\ 0 < t < \infty)$ where $\delta g = t^{-1}\, d/dt\, (g(t))$, $\delta^k(g) = \delta(\delta^{k-1}(g))$. Define h on R^n by: $h(x) = g(|x|)$, where $|x|$ is the distance from x to the origin. Then h is the Fourier transform of some f in $L^1(R^n)$."

We shall, however, give an alternative proof. Since $tb'(t)$ is slowly varying, $tb'(t)$ has constant sign for t sufficiently large by Lemma 1. Therefore $b'(t)$ has constant sign for t sufficiently large. Thus if a is sufficiently large and if $t > a$,

$$|b(t)-b(a)| = \left|\int_a^t b'(s)ds\right| = \int_a^t |b'(s)|\, ds.$$

$b(t) \to 0$ as $t \to \infty$. Hence by letting $t \to \infty$, we see b' is in L^1. It now follows from ii) and 1.53 that $F(x)$ is in L^1.

Assume for a moment we could show

1.57 $$\int_{E_k} |F_\varepsilon(x)|\, dx \leq M < \infty,$$

for some M not depending on ε. Then it would follow that for some sequence

$\varepsilon_j \to 0+$, $F_{\varepsilon_j}(x)$ would tend weakly to a measure $d\mu$. Moreover assertion ii) shows that outside of any neighborhood of the origin, $F_\varepsilon(x)$ is dominated by an L^1 function. Also by the Fourier inversion formula

$$\psi(|x|)b(|x|)\exp(-\varepsilon|x|) = \int_{E_k} \exp(-2\pi i x \cdot y)F_\varepsilon(y)dy.$$

Consequently

1.58 $$\psi(|x|)b(|x|) = \int_{E_k} \exp(-2\pi i x \cdot y)d\mu(y).$$

But we also know that the support of the singular part of $d\mu$ must be entirely concentrated at the origin. So if $d\mu$ had a singular part, $(d\mu)\hat{}$ could not tend to zero as $x \to \infty$. But by 1.58 and the hypothesis that $b(t) \to 0$ as $t \to \infty$, we see $(d\mu)\hat{} \to 0$. Therefore $d\mu$ must be absolutely continuous. Let $H(x)$ be the Radon Nikodym derivative of $d\mu$. Then $H(x)$ is in L^1;

$$\psi(|x|)b(|x|) = H\hat{}(x),$$

and $F_\varepsilon(x)$ converges to $H(x)$ almost everywhere as $\varepsilon \to 0+$. Hence $H(x) = F(x)$.

It remains to show 1.57. Using Lemma 8 instead of Lemma 4, we can obtain an equation similar to 1.25. Namely,

1.59 $$F_\varepsilon(x) = 2\pi R^{\nu+\frac{1}{2}(2-k)} \sum e(m,p,g,q)\varepsilon^m \cdot \int_0^\infty \psi^{(g)}(t)\, t^{-\nu+m+g+\frac{1}{2}k} h_p(t) J_{\nu+\frac{1}{2}(k-2)}(2\pi tR)dt,$$

with $\nu \geq 2 + \frac{1}{2}k$. The sum is extended over non-negative integral values of m, p, g, q such that $m + p + g + q = \nu$; there is however no term with $p = g = m = 0$. (This last condition is the essential difference between 1.59 and 1.25). It is trivial to dominate the terms of 1.59 with $g \geq 1$ by an L^1 function. Hence it suffices to show that

1.60 $$\int_0^\infty R^{\frac{1}{2}k}\varepsilon^m \int_0^\infty \psi(t)t^{m-\nu+\frac{1}{2}k}|h_p(t)|\exp(-\varepsilon t)|J_{\frac{1}{2}(k-2)+\nu}(2\pi tR)|\,dt\,dR < M < \infty,$$

where M does not depend on ε. Also in 1.60 at least one of p and m is greater than zero. By Fubini's Theorem, the right-hand side of 1.60 is

$$\varepsilon^m \int_0^\infty \psi(t)t^{m-\nu+\frac{1}{2}k}|h_p(t)|\exp(-\varepsilon t)\int_0^\infty R^{\frac{1}{2}k}|J_{\frac{1}{2}(k-2)+\nu}(2\pi tR)|\,dR\,dt$$

$$= \varepsilon^m \int_0^\infty \psi(t)t^{m-1}|h_p(t)|\exp(-\varepsilon t)\int_0^\infty u^{-\nu+\frac{1}{2}k}|J_{\frac{1}{2}(k-2)+\nu}(2\pi u)|\,du,$$

by a change of variables u = Rt. Since $\nu \geq 1 + \frac{1}{2}k$, Lemma 6 implies

$$\int_0^\infty u^{-\nu+\frac{1}{2}k}|J_{\frac{1}{2}(k-2)+\nu}(2\pi u)|\,du < \infty .$$

Therefore we need only show

$$\varepsilon^m \int_0^\infty \psi(t)t^{m-1}|h_p(t)|\exp(-\varepsilon t)$$

is bounded uniformly in $\varepsilon > 0$ if at least one of p and m is greater than zero. If $p \geq 1$, $t^{-1}h_p(t)$ is in L^1. (We have already remarked that $t^{-1}h_1(t) = b'(t)$ is in L^1, and by Lemma 1 $|h_p(t)| = O(|h_1(t)|)$ for $p \geq 1$ as $t \to \infty$.) Let $A = \max_{0 \leq \xi < \infty} \xi^m \exp(-\xi)$. Then

$$\varepsilon^m \int_0^\infty \psi(t)t^{m-1}|h_p(t)|\exp(-\varepsilon t)dt \leq A \int_0^\infty \psi(t)t^{-1}|h_p(t)|\,dt < \infty$$

for $p \geq 1$.

We now consider the case $m > 0$, $p = 0$.

$$\varepsilon^m \int_0^\infty \psi(t)t^{m-1}\exp(-\varepsilon t)|b(t)|\,dt$$

$$\leq (\max|b|)\varepsilon^m \int_{1/2}^\infty t^{m-1}\exp(-\varepsilon t)dt$$

$$\leq (\max|b|) \int_0^\infty t^{m-1} \exp(-t)dt$$

$$< \infty.$$

This finishes the proof of Theorem 3.

We shall now treat the other exceptional case, namely $\gamma = k + \ell$. First we need one more lemma about slowly varying functions.

LEMMA 9. Let $b(t)$ be slowly varying. Define

$$B(u) = \int_0^u \psi(t)t^{-1}b(t)dt.$$

If $\psi(t)t^{-1}b(t)$ is in $L^1[0, \infty]$, set

$$R(u) = \int_u^\infty \psi(t)t^{-1}b(t)dt.$$

Then, as $u \to \infty$

i) $|b(u)| = o\{|B(u)|\}$

if $\psi(t)t^{-1}b(t)$ is not in L^1.

ii) $|b(u)| = o\{|R(u)|\}$

if $\psi(t)t^{-1}b(t)$ is in L^1. For the proof of Lemma 9, see [25], Vol. I, p. 188, equations 2.14. (We may assume $b \geq 0$ by Lemma 1.)

THEOREM 4. Assume $b(t)$ is in S. Let $\gamma = k + \ell$, and let

$$F_\varepsilon(x) = \{\psi(|x|)|x|^{-\ell-k}b(|x|)\exp(-\varepsilon|x|)Y_\ell(x')\}^\vee .$$

Then conclusions i) and ii) of Theorem 1 hold.

At $x = 0$ there are two cases:

a. If $\psi(t)t^{-1}b(t)$ is not in $L^1[0,\infty)$, then

$$F(x)R^{-\ell} = 2(\pi)^{\ell+\frac{1}{2}k}\{\Gamma(\ell+\tfrac{1}{2}k)\}^{-1}Y_\ell(x')\int_1^{1/R} t^{-1}b(t)dt + o\Big(\int_1^{1/R} t^{-1}b(t)dt\Big).$$

In particular if $\ell = 0$ and $\gamma = k$, $F(x)$ is unbounded.

b. If $\psi(t)t^{-1}b(t)$ is in L^1, then

$$F(x)R^{-\ell} = 2(\pi)^{\ell+\frac{1}{2}k}\{\Gamma(\ell+\tfrac{1}{2}k)\}^{-1}Y_\ell(x')\int_0^{1/R}\psi(t)t^{-1}b(t)dt$$

$$+ o\left\{\int_{1/R}^{\infty}\psi(t)t^{-1}b(t)dt\right\} + O(R).$$

Also $F(x)$ is in L^1, and

$$F\hat{\ }(x) = \psi(|x|)|x|^{-\gamma}b(|x|)Y_\ell(x').$$

REMARK 1. In Case b with $\ell = 0$ and $\gamma = k$, we may extend $F(x)$ to a continuour function on all of E_k by defining

$$F(0) = 2(\pi)^{\ell+\frac{1}{2}k}\{\Gamma(\ell+\tfrac{1}{2}k)\}^{-1}\int_0^{\infty}\psi(t)t^{-1}b(t)dt;$$

(Y_0 is a constant which we take to be 1.)

Proof of Theorem 4.

The fact that conclusions i) and ii) hold is proved in the same way as i) and ii) of Theorem 1 were proved. Also the fact that $F(x)$ is in L^1, and the fact that

$$F\hat{\ }(x) = \psi(|x|)|x|^{-\gamma}b(|x|)Y_\ell(x')$$

can be shown by the method of Theorem 1. We shall therefore only consider the behavior of $F(x)$ at $x = 0$. The proof we shall give for a and b will only be valid for $k \geq 2$. However, the case $k = 1$ is so similar to the corresponding known result for series that we do not bother to give a separate proof. See [25], vol. I, p. 188, Theorem 2.15. Thus we assume $k \geq 2$. By Lemma 3,

$$F_\varepsilon(x) = 2\pi R^{\frac{1}{2}(2-k)}Y_\ell(x')\int_0^{\infty}\psi(t)t^{-\ell-\frac{1}{2}k}b(t)\exp(-\varepsilon t)J_{\frac{1}{2}(k-2)+\ell}(2\pi tR)dt.$$

Since $k \geq 2$, one can easily see

$$F(x) = \lim_{\varepsilon\to 0+} F_\varepsilon(x) = 2\pi R^{\frac{1}{2}(2-k)}Y_\ell(x')\cdot\int_0^{\infty}\psi(t)t^{-\ell-\frac{1}{2}k}b(t)J_{\frac{1}{2}(k-2)+\ell}(2\pi tR)dt.$$

We remember of course that $|J_{\frac{1}{2}(k-2)+\ell}(t)| = O(t^{-1/2})$ as $t\to\infty$ by Lemma 5. Therefore,

1.61 $$F(x)R^{-\ell} = 2\pi\, Y_\ell(x') \int_0^\infty \psi(t)t^{-1}b(t)\cdot(Rt)^{-\frac{1}{2}(k-2)-\ell} J_{\frac{1}{2}(k-2)+\ell}(2\pi tR)dt.$$

Assume first $\psi(t)t^{-1}b(t)$ is not in L^1. Then from 1.61,

1.62 $$F(x)R^{-\ell} = 2(\pi)^{\ell+\frac{1}{2}k}\{\Gamma(\ell+\tfrac{1}{2}k)\}^{-1}Y_\ell(x')\int_0^{1/R}\psi(t)t^{-1}b(t)dt$$

$$- 2\pi\, Y_\ell(x')\int_0^{1/R}\psi(t)t^{-1}b(t)\Big\{\pi^{\ell+\frac{1}{2}(k-2)}\{\Gamma(\ell+\tfrac{1}{2}k)\}^{-1}$$

$$- (Rt)^{-\frac{1}{2}(k-2)-\ell}J_{\frac{1}{2}(k-2)+\ell}(2\pi tR)\Big\}dt$$

$$+ (2\pi)Y_\ell(x')\int_{1/R}^\infty t^{-1}b(t)(Rt)^{\ell-\frac{1}{2}(k-2)}J_{\frac{1}{2}(k-2)+\ell}(2\pi tR)dt.$$

We now consider the integrals in 1.62. First

$$\left|\int_0^{1/R}\psi(t)t^{-1}b(t)\Big\{\pi^{\ell+\frac{1}{2}(k-2)}\{\Gamma(\ell+\tfrac{1}{2}k)\}^{-1} - (Rt)^{-\ell-\frac{1}{2}(k-2)}J_{\frac{1}{2}(k-2)+\ell}(2\pi tR)\Big\}dt\right|$$

$$= O\Big\{R\int_0^{1/R}\psi(t)|b(t)|dt\Big\}$$

by equation 1.7 in Lemma 5. Further,

$$\int_0^{1/R}\psi(t)|b(t)|dt = \int_0^{1/R}t^{-\frac{1}{2}}\psi(t)t^{\frac{1}{2}}|b(t)|dt = O(1) + O\Big\{R^{-\frac{1}{2}}b(1/R)\int_0^{1/R}\psi(t)t^{-\frac{1}{2}}dt\Big\}$$

$$= O(1) + O(R^{-1}\, b(1/R)),$$

where we have used the fact that $t^{1/2}|b(t)|$ is increasing for sufficiently large t. Thus

1.63 $$\left|\int_0^{1/R}\psi(t)t^{-1}b(t)\Big\{\pi^{\ell+\frac{1}{2}(k-2)}\{\Gamma(\ell+\tfrac{1}{2}k)\}^{-1} - (Rt)^{-\ell-\frac{1}{2}k}J_{\frac{1}{2}(k-2)+\ell}(2\pi tR)\Big\}dt\right|$$

$$= O(|b(1/R)|).$$

Next

$$\left|\int_{1/R}^{\infty} t^{-1}b(t)(Rt)^{-\ell-\frac{1}{2}(k-2)}J_{\frac{1}{2}(k-2)+\ell}(2\pi tR)dt\right|$$

$$\leq R^{-\frac{1}{2}(k-2)-\ell}\int_{1/R}^{\infty} t^{-1/4}(|b(t)|)\, t^{-3/4-\ell-\frac{1}{2}(k-2)}|J_{\frac{1}{2}(k-2)+\ell}(2\pi tR)|\,dt$$

$$= O\left\{R^{-\ell-\frac{1}{2}(k-2)}R^{1/4}b(1/R)\int_{1/R}^{\infty} t^{-3/4-\ell-\frac{1}{2}(k-2)}|J_{\frac{1}{2}(k-2)+\ell}(2\pi tR)|\,dt\right\}$$

since $|b(t)|t^{-1/4}$ is decreasing for large t. Now using equation 1.10 in Lemma 6, we see

$$1.64 \qquad \left|\int_{1/R}^{\infty} t^{-1}b(t)(Rt)^{-\ell-\frac{1}{2}(k-2)}J_{\frac{1}{2}(k-2)+\ell}(2\pi tR)dt\right| = O(b(1/R)).$$

Substituting 1.63 and 1.64 into 1.62, we obtain

$$F(x)R^{-\ell} = 2Y_{\ell}(x')\,\pi^{\ell+\frac{1}{2}k}\{\Gamma(\ell+\tfrac{1}{2}k)\}^{-1}\int_{0}^{1/R}\psi(t)t^{-1}b(t)dt + O(1) + O\{|b(1/R)|\}.$$

Hence a follows from Lemma 9.

We consider now the case that $\psi(t)t^{-1}b(t)$ is in L^1. From 1.61

$$F(x)R^{-\ell} = 2\pi^{\ell+\frac{1}{2}k}\{\Gamma(\ell+\tfrac{1}{2}k)\}^{-1}Y_{\ell}(x')\int_{0}^{1/R}\psi(t)t^{-1}b(t)dt + E(x).$$

Where

$$E(x) = 2\pi^{\ell+\frac{1}{2}k}\{\Gamma(\ell+\tfrac{1}{2}k)\}^{-1}Y_{\ell}(x')\int_{1/R}^{\infty} t^{-1}b(t)\;(Rt)^{-\ell-\frac{1}{2}(k-2)}J_{\ell+\frac{1}{2}(k-2)}(2\pi tR)dt$$

$$+ 2\pi\; Y_{\ell}(x')\int_{0}^{1/R}\psi(t)t^{-1}b(t)$$

$$\cdot\left\{(Rt)^{-\ell-\frac{1}{2}(k-2)}J_{\ell+\frac{1}{2}(k-2)}(2\pi tR) - \pi^{\ell+\frac{1}{2}(k-2)}\{\Gamma(\ell+\tfrac{1}{2}k)\}^{-1}\right\}dt.$$

By 1.64 and the reasoning of 1.63,

$$|E(x)| = O(|b(1/R)|) + O(R).$$

Thus statement b follows.

In Theorem 5 below, we shall show that our previous theorems may be used to calculate the asymptotic behavior of the derivatives of the $F(x)$ of Theorems 1, 2, 3, and 4. In Theorem 5 we shall use the following notation. Let $s = (s_1, s_2, \ldots, s_k)$. Then $x^s = x_1^{s_1} \ldots x_k^{s_k}$.

THEOREM 5. Assume $b(t)$ is in S, and suppose $W(x')$ is in $C^\infty(S_{k-1})$. Let γ be any real number. Take $s = (s_1, s_2, \ldots, s_k)$; s_k non-negative integers. Define

$$F_\varepsilon(x) = \{\psi(|x|)|x|^{-\gamma} b(|x|) W(x') \exp(-\varepsilon|x|)\}^\vee .$$

Then $F(x) = \lim_{\varepsilon \to 0+} F_\varepsilon(x)$ exists for $x \neq 0$. Also for $x \neq 0$

$$D^s F(x) = G(x) = \lim_{\varepsilon \to 0+} G_\varepsilon(x);$$

where

$$G_\varepsilon(x) = \{\psi(|x|)|x|^{-\gamma_1} b(|x|)\Omega(x') \exp(-\varepsilon|x|)\}^\vee .$$

$\gamma_1 = \gamma - \|s\|$. $\Omega(x') = |x|^{-\|s\|}(2\pi i x)^s \Omega(x')$. Since $W(x')$ is in $C^\infty(S^{k-1})$, so is $\Omega(x')$.

Proof of Theorem 5. If $\gamma \leq k$, Theorem 2, 3, or 4 asserts that $F(x)$ exists and is continuous for $x \neq 0$. If $k > \gamma$, the existence and continuity of $F(x)$ are obvious. Similarly, we see that $G(x)$ exists and is continuous for $x \neq 0$. It is easy to see that the weak s^{th} derivative of $F(x) = G(x)$ for x outside of any neighborhood of the origin. Since F, the ordinary derivatives of F, and G are continuous for $x \neq 0$, it follows that $D^s F = G$ in the ordinary sense for $x \neq 0$. This finishes the proof of Theorem 5.

Our next objective is to carry our results about trigonometric integrals into theorems about trigonometric series. Our main tool will be Lemma 10.

LEMMA 10. (The Poisson-summation formula).

i) Assume $G(x)$ is in $L^1(E_k)$. Then $\Sigma_{(n)}G(x+n)$ converges in $L^1(T^k)$ to a function $\widetilde{G}(x)$. If we denote $G^\wedge(x)$ by $\phi(x)$, then the Fourier series of $\widetilde{G}$ is

$$\sum_{(n)}\phi(n)\exp(2\pi in\cdot x).$$

ii) Assume $G(x)$ is continuous, $\Sigma_{(n)}G(x+n)$ converges uniformly, and $\Sigma|\phi(n)| < \infty$. Then

1.65 $$\widetilde{G}(x) = \sum_{n}\phi(n)\exp(2\pi in\cdot x).$$

For the proof of Lemma 10, see [4], p. 30, Theorems 2.41 and 2.42.

The next lemma will be used in part II of our paper as well as in part I.

LEMMA 11. Let $\Phi(x)$ be a continuous function on E_k. Assume that for every $\varepsilon > 0$, $|\Phi(x)| = O(\exp(\varepsilon|x|))$ as $x\to\infty$. Set

$$F_\varepsilon(x) = \{\exp(-\varepsilon|x|)\,\Phi(x)\}^\vee.$$

Suppose $F(x) = \lim_{\varepsilon\to 0+} F_\varepsilon(x)$ exists in the pointwise sense for $x \neq 0$, and $F(x)$ is continuous for $x \neq 0$. Furthermore, for some $\mu > 0$, assume $|F_\varepsilon(x)| = O\{|x|^{-k-\mu}\}$ uniformly in $\varepsilon \geq 0$ as $x\to\infty$. Then

i) $$f(x) = \lim_{\varepsilon\to 0+}\sum \exp(-\varepsilon n)\Phi(n)\exp(2\pi in\cdot x)$$

exists in the pointwise sense for values of x which are not lattice points.

ii) $f(x) = F(x) + E(x)$;

$E(x)$ is continuous for x in T^k.

iii) If $F(x)$ is in $L^1(E_k)$ then $f(x)$ is in $L^1(T^k)$, and the nth Fourier coefficient of f is $\Phi(n)$.

iv) If $F(x)$ is infinitely differentiable for $x \neq 0$, and if each derivative of $F(x)$ is $O(|x|^{-k-1})$ as $x\to\infty$, then $E(x)$ is in $C^\infty(T^k)$.

Proof: By the Fourier inversion formula

$$F_\varepsilon\hat{\ }(x) = \exp(-\varepsilon|x|)\,\Phi(x).$$

The hypothesis of ii) in Lemma 10 are satisfied with $G(x) = F_\varepsilon(x)$, and $\phi(x) = \exp(-\varepsilon|x|)\Phi(x)$. Therefore by equation 1.65

$$\sum \exp(-\varepsilon|n|)\Phi(n)\exp(2\pi i n\cdot x) = \sum F_\varepsilon(x+n).$$

Since $|F_\varepsilon(x)| = O\{|x|^{-k-\mu}\}$ uniformly in ε, we can take the limit as $\varepsilon \to 0+$ under the summation sign. We see that $f(x)$ exists, and

1.66 $$f(x) = F(x) + E(x)$$

with

1.67 $$E(x) = \sum_{n\neq 0} F(x+n).$$

Since $F(x) = O(|x|^{-k-1})$, and since $F(x)$ is continuous for $x \neq 0$, the series defining $E(x)$ in 1.67 is a uniformly convergent sequence of continuous functions for x in T^k. Therefore, $E(x)$ is continuous for x in T^k. This proves assertions i) and ii). Assertion iii) follows immediately from conclusion i) of Lemma 10. iv) is trivial in view of 1.67.

We shall now use Lemma 11 with

$$\Phi(x) = \psi(|x|)|x|^{-\gamma}\, b(|x|)W(x'),$$

where $W(x')$ is in $C^\infty(S_{k-1})$. Then Theorems 6 and 7 below are trivial corollaries of Theorems 1, 2, 3, 4, and 5.

THEOREM 6. Let $b(t)$ be in S, and assume $W(x')$ is in $C^\infty(S_{k-1})$. Define

$$f_\varepsilon(x) = \sum{}' |n|^{-\gamma}\, b(|n|)W(n')\exp(-\varepsilon n + 2\pi i n\cdot x).$$

Then $f(x) = \lim_{\varepsilon\to 0+} f_\varepsilon(x)$ exists if x is not a lattice point, and $f(x)$ is infinitely differentiable except for values of x which are lattice points.

THEOREM 7.

i) Let $W(x')$ satisfy the hypothesis of Theorem 2. Assume $f(x)$ is the $f(x)$ of Theorem 6. Then

$$f(x) = |x|^{\gamma-k} b(1/|x|)\widetilde{W}_{\gamma,k}(x') + o\{|x|^{\gamma-k} b(1/|x|)\} + O(1),$$

as $x \to 0$. $f(x)$ is in $L^1(T^k)$ if and only if $\psi(t)t^{-1-\gamma} b(t)$ is in $L^1[0,\infty]$. Also if $f(x)$ is in L^1,

$$f(x) \sim \sum{}' |n|^{-\gamma} b(|n|)\widetilde{W}_{\gamma,k}(n')\exp(2\pi in\cdot x).$$

$\widetilde{W}_{\gamma,k}$ is defined in equation 1.52.

ii) If $\ell + \gamma = -2m$ for some non-negative integer m, and if $W(x') = Y_\ell(x')$, then

$$f(x) = A|x|^{-k-1+\gamma} b'(1/|x|)Y_\ell(x') + o\{|x|^{-k-1+\gamma} b'(1/|x|)\} + O(1),$$

as $x \to 0$. $A = A(k,\gamma) \neq 0$.

If $\ell = \gamma = 0$ and if $b(t) \to 0$ as $t \to \infty$, $f(x)$ is in L^1, and

$$f(x) \sim \sum_{n\neq 0} b(|n|)\exp(2\pi in\cdot x).$$

iii) Assume $\gamma = k$. If

a. $\psi(t)t^{-1}b(t)$ is not in $L^1[0,\infty]$, then

$$f(x) = B\int_1^{1/|x|} t^{-1} b(t)dt + o\{\int_1^{1/|x|} t^{-1} b(t)dt\}.$$

$B = B(k) \neq 0$.

If

b. $\psi(t)t^{-1}b(t)$ is in $L^1[0,\infty]$, then $f(x)$ is extendable to a continuous function.

In both cases a and case b, $f(x)$ is in $L^1(T^k)$, and

$$f(x) \sim \sum_{n\neq 0} |n|^{-k} b(|n|)\exp(2\pi in\cdot x).$$

REMARK: Let $0 < \gamma < k$. Then

$$\lim_{\varepsilon \to 0+} \sum{}' |n|^{-\gamma} \exp(-\varepsilon n + 2\pi i n \cdot x) = \pi^{\gamma - \frac{1}{2}k} \Gamma(\tfrac{1}{2}(k-\gamma))\{\Gamma(\tfrac{1}{2}\gamma)\}^{-1} |x|^{\gamma - k} + E(x).$$

$E(x)$ is in $C^\infty(T^k)$.

The proof of Theorems 6 and 7 follow from Lemma 11 and Theorems 1-5. We make only one remark. By Theorem 5, the derivatives of an $F(x)$ of Theorems 2, 3 or 4 is another $F(x)$ in the conclusions of one of Theorems 2, 3, or 4. Hence $|D^s F(x)| = O(|x|^{-r})$ as $x \to \infty$ for any integer r. (The O depends on r and s.)

Finally we shall need a result which enables us to calculate the asymptotic behavior of the derivatives of $f(x)$.

THEOREM 8. Assume $b(t)$ is in S, and suppose $W(x')$ is in $C^\infty(S_{k-1})$. Let γ be any real number. Take $s = (s_1, \ldots, s_k)$; s_j are non-negative integers. Define

$$f_\varepsilon(x) = \sum_{n \neq 0} |n|^{-\gamma} b(|n|) W(n') \exp(-\varepsilon |n| + 2\pi i n \cdot x).$$

Then $f(x) = \lim_{\varepsilon \to 0+} f_\varepsilon(x)$ exists for $x \neq 0$ in T^k. Also for $x \neq 0$ in T^k, $D^s f(x)$ exists, and

$$D^s f(x) = g(x) = \lim_{\varepsilon \to 0+} g_\varepsilon(x);$$

where

$$g_\varepsilon(x) = \sum_{n \neq 0} |n|^{-\gamma_1} b(|n|) \Omega(n') \exp(-\varepsilon |n| + 2\pi i n \cdot x).$$

$$\gamma_1 = \gamma - \|s\|, \text{ and } \Omega(x') = |x|^{-\|s\|} (-2\pi i x)^s W(x').$$

The proof of Theorem 8 is the same as the proof of Theorem 5.

PART II

The method of part II is less elementary than the technique of part I in that complex integration will be used. Our first goal will be to prove Theorem 9 below. Methods similar to ours have been used by Berg and Mikusinski. See [1], [2], and [18]. After obtaining Theorem 9, we shall employ the Poisson summation formula to obtain an analogous theorem for trigonometric series. The one-dimensional case is considered by a different method in [13].

Here, as usual, $\psi(t)$ is in $C^{\infty}(-\infty,\infty)$, $\psi(t) = 0$ for $t \leq \frac{1}{2}$, $\psi(t) = 1$ for $t \geq 1$, and $0 \leq \psi(t) \leq 1$.

THEOREM 9. Let $0 < a < 1$. Define

$$F_{\varepsilon}(x) = \{\psi(|x|)|x|^{-b} \exp(-\varepsilon|x| - 2\pi i|x|^{a})\}^{\vee}.$$

Then $F_{\varepsilon}(x)$ depends only on R. Furthermore

i) $F(x) = \lim_{\varepsilon \to 0+} F_{\varepsilon}(x)$ exists in the pointwise sense for $x \neq 0$, and $F(x)$ is continuous for $x \neq 0$.

ii) If $b > k(1 - \frac{1}{2}a)$, $F(x)$ is continuous. If $b \leq k(1 - \frac{1}{2}a)$, and if m_0 is any non-negative integer,

2.1 $$F(x) = R^{-(k-b-\frac{1}{2}(ka))/(1-a)} \exp(i\xi_a R^{-a/(1-a)}) \cdot \sum_{m=0}^{m_0} \alpha_m R^{ma/(1-a)} + O(R^{(m_0+1)a/(1-a)}) + C(R).$$

$a_0 \neq 0$; ξ_a is real, and $\xi_a \neq 0$; $C(R)$ is a continuous function.

iii) If $b \geq k(1 - \frac{1}{2}a)$, $F_{\varepsilon}(x)$ is uniformly bounded in $\varepsilon \geq 0$. If $b < k(1 - \frac{1}{2}a)$,

$$|F_{\varepsilon}(x)| = O\{R^{-(k-b-\frac{1}{2}(ka))/(1-a)}\}$$

uniformly in ε as $R \to 0$.

iv) For any integer r, $|F_\varepsilon(x)| = O(R^{-r})$ uniformly in ε as $R \to \infty$.

v) $F(x)$ is in L^1 if and only if $b > \frac{1}{2}(ka)$. If $f(x)$ is in L^1,

$$F^\wedge(x) = \psi(|x|)|x|^{-b} \exp(2\pi i |x|^a).$$

Before we prove Theorem 9, we shall state three lemmas.

LEMMA 12. Let $\eta_1 > 0$, let $\mu \geq -\frac{1}{2}$, and let $0 < \eta_2 < \frac{1}{2}\pi$. Suppose Ω is a domain in the complex domain defined by $|z| \geq \eta_1$ and $|\arg z| \leq \eta_2$. Then $J_\mu(z)$ is analytic and single-valued in Ω. Moreover

2.2 $$J_\mu(z) = (\tfrac{1}{2}\pi z)^{-\frac{1}{2}} \exp(-iz) S_1^\mu(z) + (\tfrac{1}{2}\pi z)^{-\frac{1}{2}} \exp(iz) S_2^\mu(z).$$

The S_j^μ are analytic in Ω. Also if m_0 is any non-negative integer

2.3 $$S_j^\mu(z) = \sum_{m=0}^{m_0} \beta^j_{\mu,m} z^{-m} + O(|z|^{-m_0-1}).$$

The $\beta^j_{\mu,m}$ are complex numbers, and $\beta^j_{\mu,0} \not\equiv 0$ for all μ and $j = 1, 2$. The O is uniform in Ω.

The proof of Lemma 12 is in [24], p. 196.

LEMMA 13. Let A and B be points in the complex plane joined by a line segment L. Suppose $g(z)$ and $h(z)$ are functions analytic in a connected domain D containing L. Define

$$\mathscr{F}(t) = \int_L g(z) \exp\{t h(z)\} dz.$$

Assume there exists a point ζ on L such that the following four conditions are satisfied:

i) $h'(\zeta) = 0$.

ii) $h''(\zeta) \neq 0$.

iii) $h''(\zeta)(\zeta - z)^2$ is real and negative for z in L, except at $z = \zeta$.

iv) $\mathscr{R}\!\ell\, h(z) - \mathscr{R}\!\ell\, h(\zeta) \leq \frac{1}{4} h''(\zeta)(\zeta - z)^2$ for z on L.

Then, as $t \to \infty$,

$$\mathscr{F}(t) = t^{-\frac{1}{2}} \exp(th(\zeta)) \left\{ \sum_{p=0}^{p_0} d_p t^{-p} + O(t^{-p_0-1}) \right\}$$

for any non-negative integer p_0.

$$|d_0| = (2\pi)^{\frac{1}{2}} |h''(\zeta)|^{\frac{1}{2}} |g(\zeta)|.$$

Lemma 13 is proved in section 5.7, p. 87 of [9]. De Bruijn does not state a specific theorem, but our Lemma 13 is proved by his reasoning. Our A is his A_2, and our B is his B_2.

LEMMA 14. Let g, h, A, B, L, and ζ satisfy the hypothesis of Lemma 13. Suppose also $\mathscr{Re}\, z \geq 0$ on L, and that $\delta(t) \geq 0$ for $t \geq 0$. Set

$$\mathscr{F}_{\delta(t)} = \int_L |g(z)\exp\{-\delta(t)z + th(z)\}|\,|dz|.$$

Then as $t \to \infty$,

$$\mathscr{F}_{\delta(t)} = O\{t^{-\frac{1}{2}} \exp\{t\,\mathscr{Re}(h(\zeta))\}\}.$$

The capital O does not depend on $\delta(t)$.

<u>Proof of Lemma 14.</u> Clearly

$$\mathscr{F}_\delta(t) = O\left\{ \int_L \exp\{t\,\mathscr{Re}(h(z))\}|dz| \right\}$$

$$= O\left\{ \exp\{t\,\mathscr{Re}(h(\zeta))\} \int_L \exp\{t(\mathscr{Re}(h(z)) - \mathscr{Re}(h(\zeta)))\}|dz| \right\}$$

$$= O\left\{ \exp\{t\,\mathscr{Re}(h(\zeta))\} \int_L \exp\{(1/4)th''(\zeta)(\zeta - z)^2\}|dz| \right\},$$

since $\mathscr{Re} h(z) - \mathscr{Re}\, h(\zeta) \leq (1/4)h''(\zeta)(\zeta-z)^2$. We parameterize L by $z = \zeta + \alpha \exp(i\lambda)$, with α and λ real. Then

$$|\mathscr{F}_\delta(t)| = O\left\{ \exp\{t\,\mathscr{Re}(h(\zeta))\} \int_{-\infty}^{\infty} \exp\{-(1/4)\alpha^2 t |h''(\zeta)|\}\,d\alpha \right\},$$

since $h''(\zeta)(\zeta-z)^2 < 0$ (except at $z = \zeta$). But

$$\int_{-\infty}^{\infty} \exp\{-(1/4)\alpha^2 t|h''(\zeta)|\}d\alpha = O(t^{-\frac{1}{2}}).$$

Therefore,

$$\mathscr{F}_\delta(t) = O\{t^{-\frac{1}{2}}\exp\{t\,\mathscr{R}\ell(h(\zeta))\}\}.$$

Proof of Theorem 9. By Lemma 3 of part I

2.4 $$F_\varepsilon(x) = 2\pi R^{\frac{1}{2}(2-k)} \int_0^\infty \psi(u)u^{-b+\frac{1}{2}k} \exp(2\pi i u^a - \varepsilon u)\, J_{\frac{1}{2}(k-2)}(2\pi R u)\, du.$$

Define

2.5 $$G_\varepsilon(x) = 2\pi R^{\frac{1}{2}(2-k)}\int_1^\infty u^{-b+\frac{1}{2}k}\exp(2\pi i u^a - \varepsilon u)J_{\frac{1}{2}(k-2)}(2\pi R u)du,$$

and let

2.6 $$E_{1,\varepsilon}(x) = 2\pi R^{\frac{1}{2}(2-k)} \int_0^1 \psi(u)u^{-b+\frac{1}{2}k}\exp(2\pi i u^a - \varepsilon u)J_{\frac{1}{2}(k-2)}(2\pi R u)du.$$

Then

2.7 $$F_\varepsilon(x) = G_\varepsilon(x) + E_{1,\varepsilon}(x).$$

By Lemma 5, equation 1.7, one can easily see that $E_1(x) = \lim_{\varepsilon \to 0+} E_{1,\varepsilon}(x)$ exists for every x and is continuous. Moreover it is clear that $E_{1,\varepsilon}(x)$ is bounded uniformly in ε.

In the integral of 2.5, we make a change of variables $s = uR^{1/(1-a)}$. We then have

2.8 $$G_\varepsilon(x) = 2\pi R^{(b-k)/(1-a)}t^{1-\frac{1}{2}k} \int_{t^{-1/a}}^\infty s^{-b+\frac{1}{2}k} \cdot \exp(2\pi i t s^a - \delta s)J_{\frac{1}{2}(k-2)}(2\pi s t)ds.$$

Here

2.9 $$t = R^{-a/(1-a)},$$

and

2.10 $$\delta = \varepsilon\, R^{-1/(1-a)}.$$

As $R \to 0$, $t \to \infty$; and δ is a non-negative function of t.

We shall consider the integral in 2.8 as an integral of a function of a complex variable s along a half line $\arg s = 0$, $\mathcal{R}\ell\, s \geq t^{-(1/a)}$. We shall then deform the contour of integration so that we may use the method of steepest descents. We shall write $s = \sigma + i\tau$, and $s = \rho \exp(i\phi)$. As a preliminary step we recall from Lemma 12 that

2.11 $$J_{\frac{1}{2}(k-2)}(2\pi st) = (\pi(st)^{\frac{1}{2}})^{-1} \exp(-2\pi ist)S_1^{\frac{1}{2}(k-2)}(2\pi st) + (\pi(st)^{\frac{1}{2}})^{-1} \exp(2\pi ist)S_2^{\frac{1}{2}(k-2)}(2\pi st).$$

Now an easy consequence of Cauchy's Theorem is that

2.12 $$G_\varepsilon(x) = H_\varepsilon(x) + E_{\varepsilon,2}(x) + E_{\varepsilon,3}(x).$$

2.13 $$H_\varepsilon(x) = 2R^{(b-k)/(1-a)}t^{\frac{1}{2}(1-k)} \cdot \int_{c_1} s^{-b+\frac{1}{2}(k-1)} \exp\{th_1(s) - \delta s\} S_1^{\frac{1}{2}(k-2)}(2\pi st)ds.$$

2.14 $$h_1(s) = 2\pi is^a - 2\pi is.$$

2.15 $$E_{\varepsilon,2}(x) = 2R^{(b-k)/(1-a)}t^{\frac{1}{2}(1-k)} \int_{c_2} s^{-b+\frac{1}{2}(k-1)} \exp\{th_2(s) - \delta s\} S_2^{\frac{1}{2}(k-2)}(2\pi st)ds.$$

2.16 $$h_2(s) = 2\pi is^a + 2\pi is.$$

2.17 $$E_{\varepsilon,3}(x) = 2\pi R^{(b-k)/(1-a)}t^{1-\frac{1}{2}k} \int_{c_3} s^{-b+\frac{1}{2}k} \exp(2\pi its^a - \delta s)J_{\frac{1}{2}(k-2)}(2\pi st)ds.$$

c_1, c_2, and c_3 are contours in the complex plane which we shall now describe. We choose our contour c_1 in such a way that $\mathcal{R}\ell\, h_1$ is negative on c_1 except at one point, where $\mathcal{R}\ell\, h_1 = 0$. We choose c_2 so that $\mathcal{R}\ell\, h_2$ is negative on c_2. The contours c_1, c_2, and c_3 will depend on an angle θ, $0 < \theta < \pi/4$ which we shall describe later. (See Figure 1.)

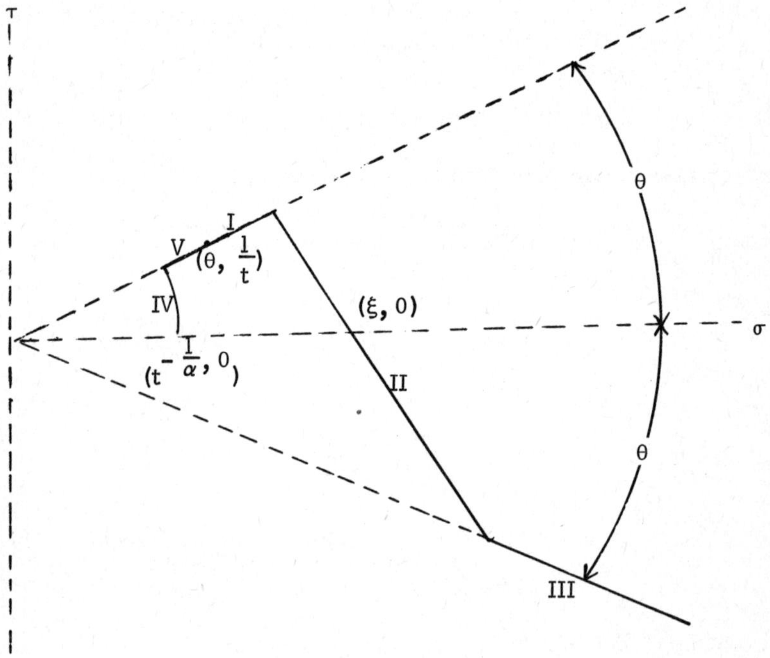

FIGURE 1

The contour c_1 consists of three parts — I, II, and III. (Remember $s = \rho \exp(i\phi)$, and $s = \sigma + i\tau$). We set $\xi = a^{1/(1-a)}$. I is the line segment defined by

$$\phi = \theta$$

and

$$t^{-1} \leq \rho \leq \tfrac{1}{2}\xi 2^{\frac{1}{2}} \csc(\theta + (\pi/4)).$$

II will be a line segment intersecting I at the point with polar coordinates

$$(\tfrac{1}{2}\xi 2^{\frac{1}{2}} \csc(\theta + (\pi/4)), \theta).$$

II is that portion of the line $\sigma = -\tau + \xi$ cut off by the rays $\phi = \theta$ and $\phi = -\theta$.

III is that portion of the ray $\phi = -\theta$ defined by

$$\rho \geq \tfrac{1}{2}2^{\frac{1}{2}}\xi \csc((\pi/4) - \theta).$$

III meets II at a point with polar coordinates

$$(\tfrac{1}{2}\sqrt{2}\xi \csc((\pi/4) - \theta), -\theta).$$

We remark that the chief contribution to $G_\varepsilon(x)$ will come from the piece of the contour designated by II.

The contour c_2 consists of that part of the ray $\phi = \theta$ defined by $\rho \geq t^{-1}$.

The contour c_3 consists of the portions IV and V in Figure 1. IV is the arc of the circle $\rho = t^{-1/a}$ contained between $\phi = 0$ and $\phi = \theta$.

V is the segment of the line $\phi = \theta$ for which

$$t^{-1/a} \leq \rho \leq t^{-1}.$$

We wish to choose θ in such a way that conditions i), ii), iii), and iv) of Lemma 13 will be satisfied by our $h_1(s)$ with the line segment $L = \text{II}$, and the point $\zeta = (\xi, 0) = (a^{1/(1-a)}, 0)$. Furthermore we wish to be sure that $\mathcal{R}\ell\, h_1$ is negative on I and III.

In our following discussion we shall always assume $-\frac{1}{2}\pi \leq \phi \leq \frac{1}{2}\pi$. That is, we are considering points s in the right half plane. One easily sees that

2.18 $$(2\pi)^{-1}\mathscr{R}\ell(h_1(s)) = -\rho^a \sin a\phi + \rho \sin \phi.$$

Clearly, then, $\mathscr{R}\ell(h_1(s)) = 0$ if $\rho = 0$ or $\sin\phi = 0$; i.e., if s is real. If $\rho \neq 0$ and if $\sin\phi \neq 0$, $\mathscr{R}\ell(h_1(s)) = 0$ if and only if

2.19 $$\rho = (\sin a\phi \csc \phi)^{1/(1-a)}.$$

We extend the curve defined by 2.19 to $\phi = 0$ by continuity. I.e., we set $\rho = \xi = a^{1/(1-a)}$ at $\phi = 0$. We call this curve Λ. (We are assuming Λ to be defined only for $-\frac{1}{2}\pi \leq \phi \leq \frac{1}{2}\pi$.) Λ intersects the real axis only at $\phi = 0$. I.e., $\rho = a^{1/(1-a)}$. Λ meets the imaginary axis at $\tau = \pm(\sin\frac{1}{2}a\pi)^{1/(1-a)}$. Thus the curve Λ together with the real axis divides the half plane $\sigma \geq 0$ into four parts $\mathscr{R}_1$, $\mathscr{R}_2$, $\mathscr{R}_3$, and $\mathscr{R}_4$.

$$\mathscr{R}_1 = \{s \mid \tau > 0 \text{ and } \rho < (\sin a\phi \csc\phi)^{1/(1-a)}\}$$

$$\mathscr{R}_2 = \{s \mid \tau < 0 \text{ and } \rho < (\sin a\phi \csc\phi)^{1/(1-a)}\}$$

$$\mathscr{R}_3 = \{s \mid \tau > 0 \text{ and } \rho > (\sin a\phi \csc\phi)^{1/(1-a)}\}$$

$$\mathscr{R}_4 = \{s \mid \tau < 0 \text{ and } \rho > (\sin a\phi \csc\phi)^{1/(1-a)}\}.$$

Certainly $\mathscr{R}\ell h_1$ has constant sign on each $\mathscr{R}_j$, $j = 1, 2, 3$, and 4. By considering very small values of ρ, one can show $\mathscr{R}\ell h_1 < 0$ in $\mathscr{R}_1$ and $\mathscr{R}\ell h_1 > 0$ in $\mathscr{R}_2$. By looking at very large values of ρ, we can see that $\mathscr{R}\ell\, h_1 > 0$ in $\mathscr{R}_3$ and $\mathscr{R}\ell\, h_1 < 0$ in $\mathscr{R}_4$. One can also show that Λ has a vertical tangent at $\rho = a^{1/(1-a)}$, $\phi = 0$. Therefore, if we take θ sufficiently small, every point of II except the point $(\xi, 0) = (a^{1/(1-a)}, 0)$ will lie in either $\mathscr{R}_1$ or $\mathscr{R}_4$. Therefore $\mathscr{R}\ell h_1$ will be negative on all of II except at $(\xi, 0)$ where $\mathscr{R}\ell\, h_1$ is zero. One can also see by a straightforward calculation that conditions i), ii), and iii) of Lemma 13 are satisfied by $h_1(s)$ with $\zeta = (\xi, 0)$ and $L = $ II.

We now consider condition iv). $h_1' = 0$ at $(\xi, 0)$. Therefore, taking $\zeta = (\xi, 0)$

$$\mathscr{R}\ell h_1(s) - \mathscr{R}\ell\, h_1(\zeta) = \mathscr{R}\ell\, \tfrac{1}{2}h_1''(\zeta)(\zeta - s)^2 + O(|\zeta - s|^3).$$

Furthermore $\mathscr{R}\ell\, h_1(s) < \mathscr{R}\ell\, h_1(\zeta)$ for $s \neq \zeta$, s in II. Also $\mathscr{R}\ell\, h_1''(\zeta)(\zeta-s)^2$ is negative for $s \neq \zeta$, s in II. Therefore, for $|\zeta - s|$ sufficiently small, we have condition iv) of Lemma 13. But we can be sure that for s in II, $|\zeta-s|$ is arbitrarily small by taking θ sufficiently small. Thus we are sure that h_1 satisfies conditions i), ii), iii), and iv) of Lemma 13 with L = II and $\zeta = (\xi, 0)$ if θ is sufficiently small.

If (ρ_0, ϕ_0) is any point in $\mathscr{R}_1$ and if $0 < \rho < \rho_0$, the point (ρ, ϕ_0) is in $\mathscr{R}_1$. Hence I is contained in $\mathscr{R}_1$, and $\mathscr{R}\ell\, h_1$ is negative on I. Similarly $\mathscr{R}\ell\, h_1$ is negative on III.

Thus we have motivated our choice of the contour c_1. However, we still must make a few remarks to obtain an asymptotic expansion for $H_\varepsilon(x)$ because, for example, the argument of the function $S_1^{\frac{1}{2}(k-2)}$ depends on t. We shall write $H_{\varepsilon, I}(x)$, $H_{\varepsilon, II}(x)$, and $H_{\varepsilon, III}(x)$ to denote the contribution to $H_\varepsilon(x)$ from the contours I, II, and III respectively. The major contribution will come from $H_{\varepsilon, II}(x)$, which we shall investigate first. As $\varepsilon \to 0$, $\delta \to 0$ by 2.10. Therefore by equation 2.13 we see that for $R \neq 0$, $H_{\varepsilon, II}(x)$ tends pointwise to a limit $H_{II}(x)$ as $\varepsilon \to 0$. Moreover,

2.20 $$H_{II}(x) = 2R^{(b-k)/(1-a)}\, t^{\frac{1}{2}(1-k)} \int_{II} s^{-b+\frac{1}{2}(k-1)} \exp\{t h_1(s)\}\, S_1^{\frac{1}{2}(k-2)}(2\pi s t)\, ds.$$

Using equation 2.3 of Lemma 12, we observe

2.21 $$H_{II}(x) = 2R^{(b-k)/(1-a)}\, t^{\frac{1}{2}(1-k)}$$
$$\cdot \left\{ \sum_{m=0}^{m_0} (2\pi)^{-m} \beta_{\frac{1}{2}(k-2),\, m}\, t^{-m} \int_{II} s^{-b+\frac{1}{2}(k-1)-m} \exp\{t h_1(s)\}\, ds \right\} + E_4(x);$$

and

2.22 $$|E_4(x)| = O\left\{ R^{(b-k)/(1-a)}\, t^{-m_0-\frac{1}{2}(k+1)} \cdot \int_{II} s^{-b-m_0-1+\frac{1}{2}(k-1)} |\exp t h_1(s)|\, ds \right\}.$$

We now apply Lemma 13 to each term of the summand in 2.21. We obtain,

2.23 $$H_{II}(x) = R^{(b-k)/(1-a)} t^{\frac{1}{2}(1-k)}$$

$$\cdot \left\{ \sum_{m=0}^{m_0} \alpha_m t^{-m-\frac{1}{2}} \exp(th_1(\xi)) + O(t^{-m_0-(3/2)}) \right\} + E_4(x).$$

(We have taken $h(z)$ to be $h_1(z)$, and we have set $g(s) = (2\pi)^{-m} s^{-b-\frac{1}{2}(k-1)}$.) Also from 2.14, we see that

2.24 $$\xi_a = -ih_1(\xi) = 2\pi a^{a/(1-a)} - 2\pi a^{1/(1-a)}.$$

Thus ξ_a is real and non-zero.

By Lemma 14,

2.25 $$E_4(x) = O\left\{ R^{(b-k)/(1-a)} t^{-m_0-2+\frac{1}{2}k} \right\}.$$

Also by 2.9, $t = R^{-a/(1-a)}$. Hence using 2.23, 2.24, and 2.25, we find

2.26 $$H_{II}(x) = R^{-(k-b-\frac{1}{2}(ka))/(1-a)} \exp(i\xi_a R^{-a/(1-a)})$$

$$\cdot \left\{ \sum_{m=0}^{m_0} \alpha_m R^{ma/(1-a)} + O(R^{(m_0+1)a/(1-a)}) \right\}.$$

Also by equation 2.20 it is easy to see that $H_{II}(x)$ is continuous for $x \neq 0$. A similar proof (starting from equation 2.13 and using Lemma 14 rather than Lemma 13) shows

2.27 $$|H_{\varepsilon, II}(x)| = O(R^{-(k-b-\frac{1}{2}ka)/(1-a)}).$$

We shall now consider $H_{\varepsilon, I}(x)$, $H_{\varepsilon, III}(x)$, $E_{\varepsilon, 2}(x)$, and $E_{\varepsilon, 3}(x)$. We shall show that each of these functions satisfies the following conditions:

2.28 Each is bounded uniformly in $\varepsilon \geq 0$.

2.29 For $x \neq 0$, each tends pointwise to a limit as $\varepsilon \to 0+$.

2.30 Each of these limit functions will be continuous for $x \neq 0$.

2.31 Each of these limit functions will tend to a limit as $x \to 0$, (i.e., as

$t \to \infty$). In view of 2.30 each of the limit functions of 2.29 will then be extendable to a continuous function.) By virtue of 2.26 and 2.27, this will conclude the proof of parts i), ii), and iii) of Theorem 9.

We consider first $H_{\varepsilon, I}(x)$. One can easily verify that on I, $\mathscr{R}\ell\, h_1(s)$ is a decreasing function of ρ for

$$t^{-1} \leq \rho \leq (a \sin a\theta \csc \theta)^{1/(1-a)}.$$

Also $\mathscr{R}\ell\, h_1(s)$ is continuous and negative on I. Therefore there exists an $\eta_1 > 0$ such that $\mathscr{R}\ell\, h_1(s) \leq -\eta_1$ for s in I and

$$\rho \geq (a \sin a\theta \csc \theta)^{1/(1-a)}.$$

Hence, using 2.18 we see

$$t\, \mathscr{R}\ell\, h_1(s) \leq \max \{-2\pi t^{1-a} \sin a\theta + \sin \theta, \ -\eta_1 t\}.$$

Hence as $t \to \infty$, (i.e., as $R \to 0$)

$$\exp \{t\, \mathscr{R}\ell\, h_1(s)\} < \eta_2 \exp(-\eta_2\, t^{1-a})$$

for some $\eta_2 > 0$. It is therefore a trivial matter to verify conditions 2.28, 2.29, and 2.31 for $H_{\varepsilon, I}(x)$. (In view of the fact that δ is a positive multiple of ε. The limit of $H_I(x)$ as $t \to \infty$ is of course zero.) 2.30 follows immediately also since the integrand is continuous on I for $t < \infty$.

We next look at $H_{\varepsilon, 3}(x)$. It is clear from 2.18, that for large ρ, say $\rho > \eta_3$,

$$\mathscr{R}\ell\, h_1(s) < -\rho\pi |\sin \phi|$$

for s on III. We also know that $\mathscr{R}\ell\, h_1(s)$ is negative on III, hence there is an $\eta_4 > 0$ such that $\mathscr{R}\ell\, h_1(s) < -\eta_4$ for s on III. Hence it is very easy to show that $H_{\varepsilon, III}$ satisfies 2.28, 2.29, 2.30, and 2.31. (Again $\lim_{t \to \infty} H_{III}(t) = 0$).

Considerations similar to those for $H_{\varepsilon, I}$ and $H_{\varepsilon, III}$ apply to $E_{\varepsilon, 2}$; and one can easily demonstrate the fact that $E_{\varepsilon, 2}$ satisfies 2.28, 2.29, 2.30, and 2.31.

We now turn to $E_{\varepsilon,3}(x)$. First we make a change of variables $z = st^{1/a}$. We then have

$$E_{\varepsilon,3}(x) = 2\pi R^{\frac{1}{2}(2-k)} \int_{C_3'} z^{-b+\frac{1}{2}k} \exp\{2\pi i z^a - \varepsilon z\} J_{\frac{1}{2}(k-2)}(2\pi z R)dz.$$

C_3' consists of two parts IV' and V'.

$$\text{IV}' = \{z \mid |z| = 1 \text{ and } 0 \leq \arg z \leq \theta\};$$

$$\text{V}' = \{z \mid \arg z = \theta \text{ and } 1 \leq |z| \leq R^{-1}\}.$$

Now

$$|\exp\{2\pi i z^a\}| \leq \exp\{-|z|^a \sin\theta\}$$

on V', and IV' is a bounded contour. Therefore one can easily see that $E_{\varepsilon,3}(x)$ satisfies conditions 2.28, 2.29, and 2.30. Also by Lemma 5 of part I,

$$\lim_{R\to 0} \{(Rz)^{\frac{1}{2}(2-k)} J_{\frac{1}{2}(k-2)}(2\pi Rz)\}$$

exists, and the convergence is bounded for $|Rz| \leq 1$. Therefore one can easily see that $E_{\varepsilon,3}(x)$ satisfies 2.31. We have now completed the proof of parts i), ii), and iii) of Theorem 9.

We turn now to part iv). $a < 1$. Therefore one can easily see that if $r > \frac{1}{2}(k-b)/(1-a)$, then

2.32 $$\int_{E_k} \left|\Delta^r\left\{\psi(|x|)|x|^{-b} \exp\{2\pi i|x|^a - \varepsilon|x|\}\right\}\right| dx \leq M < \infty.$$

M does not depend on ε. We remark that this is the crucial point at which one needs the assumption $a < 1$. Now let $q(x)$ be in $C^\infty(E_k)$. Assume $q(x)$ and all of its derivatives are in L^1. Then

2.33 $$q^\vee = (-4\pi|x|^2)^{-1}\{\Delta q\}^\vee.$$

(See [4], p. 45, equation 2.8.9). Now assertion iv) of Theorem 9 follows by 2.32 and iterated use of 2.33 if we set

$$q(x) = \psi(|x|)|x|^{-b} \exp(2\pi i|x|^a - \varepsilon|x|).$$

Finally we consider v). It is clear from ii) and iv) that $F(x)$ is in L^1 if and only if $b > \frac{1}{2}ka$. However using ii) and iv), we see that $F_\varepsilon(x)$ is dominated by an L^1 function if $b > \frac{1}{2}ka$. Now part v) is easy.

We now state a theorem for trigonometric series which is analogous to Theorem 9.

THEOREM 10. Let $0 < a < 1$. Define

$$f_\varepsilon(x) = \sum{}' |n|^{-b} \exp(-\varepsilon|n| + 2\pi i|n|^a + 2\pi i n\cdot x).$$

Then $f(x) = \lim_{\varepsilon\to 0+} f_\varepsilon(x)$ exists in the pointwise sense for x not a lattice point. For $x \neq 0$, $f(x)$ is continuous in T^k. If $b > k(1 - \frac{1}{2}a)$, $f(x)$ is extendable to a function continuous on all of E_k.

If $b \leq k(1 - \frac{1}{2}a)$,

$$f(x) = R^{-(k-b-\frac{1}{2}ak)/(1-a)} \exp\{i\xi_a R^{-1/(1-a)}\}$$

$$\cdot \left\{\sum_{m=0}^{m_0} \alpha_m R^{ma/(1-a)} + O(R^{(m_0+1)a/(1-a)})\right\} + c(x).$$

ξ_a is real and not zero, $\alpha_0 \neq 0$, and $c(x)$ is continuous in T^k. $f(x)$ is in $L^1(T^k)$ if and only if $b > \frac{1}{2}ak$. If $b > \frac{1}{2}ak$,

$$f(x) \sim \sum{}' |n|^{-b} \exp(2\pi i|n|^a + 2\pi i n\cdot x).$$

Theorem 10 follows from Theorem 9, and Lemma 11 of part I.

PART III

The simplest type of series we shall discuss in part III is

$$\sum{}' |n|^{-b} \exp\{i|n| \log|n| + 2\pi i n \cdot x\}.$$

The one-dimensional case is treated in [16], pp. 100-102. Also see [25], vol. I, pp. 197-200. As usual we shall deal with analogous trigonometric integrals first. Hence we introduce the trigonometric integral

$$F_\varepsilon(x) = \int_{E_k} \psi(|y|)|y|^{-b} \exp\{-\varepsilon|y| + i|y|\log|y| + 2\pi i y \cdot x\} dy.$$

In parts I and II the behavior of $F_\varepsilon(x)$ at ∞ was very easy to analyze. Here, however, the behavior of $F_\varepsilon(x)$ will not be trivial at infinity because the derivatives of $t^{-1}\psi(t)\exp(it \log t)$ do not behave well at infinity. We shall get to this difficulty later. We must first show that $\lim_{\varepsilon \to 0+} F_\varepsilon(x)$ exists. We therefore begin with

THEOREM 11. Let $Y_\ell(x')$ be a homogeneous spherical harmonic of degree ℓ. Assume $d > 0$. Define

$$F_\varepsilon(x) = \{\psi(|x| - 1)|x|^{-b}\log^{-c}|x| Y_\ell(x')\exp(i|x|\log^d|x|)\}^{\vee}.$$

$F(x) = \lim_{\varepsilon \to 0+} F_\varepsilon(x)$ exists in the pointwise sense. Furthermore $F(x)$ is continuous.

Proof: By Lemma 3

3.1 $$F_\varepsilon(x) = 2\pi(-i)^\ell Y_\ell(x') R^{\frac{1}{2}(2-k)} \int_0^\infty \psi(s-1) s^{-b+\frac{1}{2}k}$$

$$\cdot \log^{-c} s \exp\{-\varepsilon s + i s \log^d s\} J_{\frac{1}{2}(k-2)+\ell}(2\pi R s) ds.$$

$$F_\varepsilon(x) = E_\varepsilon(x) + 2\pi(-i)^\ell Y_\ell(x')G_\varepsilon(R).$$

$$E_\varepsilon(x) = 2\pi(-i)^\ell Y_\ell(x')R^{\frac{1}{2}(2-k)}\int_0^\infty \psi(s-1)s^{-b+\frac{1}{2}k}$$

$$\cdot \log^{-c} s \exp(-\varepsilon s + i s \log^d s) J_{\frac{1}{2}(k-2)+\ell}(2\pi Rs)\, ds.$$

$$G_\varepsilon(R) = R^{\frac{1}{2}(2-k)}\int_e^\infty s^{-b+\frac{1}{2}k} \log^{-c} s \cdot \exp(-\varepsilon s + i\ s \log^d s) J_{\frac{1}{2}(k-2)+\ell}(2\pi Rs)ds.$$

It is easy to see that $\lim_{\varepsilon \to 0+} E_\varepsilon(x)$ exists and is continuous. (Because $E_\varepsilon(x)$ is the εth Abel mean of the Fourier transform of an integrable function with compact support.)

Therefore, to prove Theorem 11, it suffices to demonstrate the following fact about $G_\varepsilon(R)$:

i) $G(R) = \lim_{\varepsilon \to 0+} G_\varepsilon(R)$ exists, and the convergence is uniform for all R in any bounded closed interval of non-negative real numbers. (Since $Y_{n,\ell}(x')G_\varepsilon(R)$ is a continuous function of x for any $\varepsilon > 0$ and since $Y_{n,\ell}$ is bounded.)

In order to prove that $G_\varepsilon(R)$ satisfies condition i), we shall use complex integration. That is, we shall consider the integral for $G_\varepsilon(R)$ as a complex integral along a part of the real axis and deform the contour of integration to a new contour Λ. Λ consists of two parts I and II. I is the arc of the circle $|s| = e$ defined by $0 \leq \arg s \leq \pi/4$. II is that portion of the ray $\arg s = \pi/4$ for which $|s| \geq e$. Since z^d is a multiple valued function, we must be sure that we never use the expression z^d when $|\arg z| \geq \pi$. But z may be of the form $\log s$. Hence we must be sure that

$$|\arg \log s| < \pi$$

in a simply connected region containing Λ and the portion of the real axis $s \geq e$. But on Λ, the relevant portion of the real axis, and in the region between them, $|\mathscr{R}\ell \log s| = \log|s| \geq 1$. Hence one easily sees $|\arg \log s| < \pi$ in a simply connected region of the desired type.

Let $s = \sigma + i\tau$. Then for $0 \leq \arg s \leq 3\pi/8$,

3.2 $$|\exp(i s \log^d s)| \leq A \exp\{-A\tau \log^d \tau\}.$$

(Here and in the following discussion, A is a positive constant not depending on R or s.) Also by Lemma 12,

3.3 $$|J_{\frac{1}{2}(k-2)+\ell}(2\pi Rs)| \leq A(R|s|)^{-\frac{1}{2}}\exp\{2\pi R\tau\}$$

for $R|s| > 1$.

Using the estimates 3.2 and 3.3, we see that for any $\varepsilon > 0$, we may employ Cauchy's theorem to move the contour of integration to Λ. Once the contour is moved to Λ, it is not difficult to prove $\lim_{\substack{\varepsilon \to 0+ \\ \varepsilon_1 \to 0+}} G_\varepsilon(R) - G_{\varepsilon_1}(R) = 0$ uniformly for R in any compact interval, and hence that $G_\varepsilon(R)$ converges uniformly, for R in any compact interval. On that portion of Λ for which $R|s| \geq 1$, one uses the estimates 3.2 and 3.3 together with the Dominated Convergence Theorem. In $R|s| \leq 1$, one employs 3.2, the Dominated Convergence Theorem, and the fact that $(sR)^{\frac{1}{2}(2-k)} J_{\frac{1}{2}(k-2)}(2\pi Rs)$ is uniformly bounded in $|Rs| \leq 1$. (See Lemma 5.) This completes the proof of Theorem 11.

Before we study the behavior of $F_\varepsilon(x)$ at infinity, we need two more lemmas.

LEMMA 15.

$$\frac{d}{dz}\left\{(\tfrac{1}{2}\pi z)^{-\frac{1}{2}} \exp(-iz) S_1^\mu(z)\right\} = \gamma^1_{1,\mu}(2\pi z)^{-\frac{1}{2}}\exp(-iz)S_1^{\mu+1}(z)$$
$$+ \gamma^1_{2,\mu}(2\pi z)^{-\frac{1}{2}} \exp(-iz) S_1^{\mu-1}(z).$$

Also

$$\frac{d}{dz}\left\{(\tfrac{1}{2}\pi z)^{-\frac{1}{2}}\exp(iz)S_2^\mu(z)\right\}$$

$$= \gamma^2_{1,\mu}(2\pi z)^{-\frac{1}{2}}\exp(iz)S_2^{\mu+1}(z) + \gamma^2_{2,\mu}(2\pi z)^{-\frac{1}{2}}\exp(iz)S_2^{\mu-1}(z).$$

The S_j^μ are the S_j^μ of Lemma 12, and the $\gamma^\mu_{i,j}$ are complex numbers of absolute value 1.

Lemma 15 is a consequence of the following two facts:

i) $(2/(\pi z))^{\frac{1}{2}} S_1^{\mu}(z)\exp(-iz)$ and $(2/(\pi z))^{\frac{1}{2}} S_2^{\mu}(z)\exp(iz)$ are the Bessel functions of the third kind, i.e., H_μ^1 and H_μ^2 respectively in the notation of [24]. (In [24] compare equations 3 and 4, p. 198 with equation 1, p. 199.)

ii) $2\frac{d}{dz} H_\mu^j(z) = H_{\mu-1}^j(z) - H_{\mu+1}^j(z)$ for $j = 1, 2$. (See equation 4, p. 174 in [24].)

LEMMA 16. Let ν be a positive integer. Then

$$\int_0^\infty \psi(y-1)y^{-b+\frac{1}{2}k}\log^{-c} y\exp(-\varepsilon y + iy\log^d y)J_{\frac{1}{2}(k-2)+\ell}(2\pi yR)\,dy = \mathscr{M}_\varepsilon(R) + \mathscr{E}_\varepsilon(R).$$

$$\mathscr{M}_\varepsilon(R) = (-2\pi iR)^\nu \int_0^\infty \psi(y-1)y^{-b+\frac{1}{2}k}\log^{-c} y$$

$$\cdot (\log^d y + d\log^{d-1} y)^\nu \exp(-\varepsilon y + iy\log^d y)J_{\frac{1}{2}(k-2)+\ell+\nu}(2\pi Ry)\,dy$$

$$= (-2\pi iR)^\nu \sum_{r=0}^{\nu} \{\nu!/(r!(\nu-r)!)\}d^{\nu-r} \cdot \int_0^\infty \psi(y-1)y^{-b+\frac{1}{2}k}(\log y)^{\{-c+d\nu-\nu+r\}}$$

$$\cdot \exp(-\varepsilon y + iy\log^d y)J_{\frac{1}{2}(k-2)+\ell+\nu}(2\pi Ry)\,dy.$$

$$\mathscr{E}_\varepsilon(t) = R^{-\nu}\sum a(m,p,j,q,r,\lambda)\varepsilon^m \int_0^\infty \psi^{(p)}(y-1)y^{-b-j-q-\lambda+\frac{1}{2}k}(\log y)^{-c+\lambda}$$

$$\cdot y^q \frac{d^q}{dy^q}(\log^d y + d\log^{d-1} y)^r \cdot \exp\{-\varepsilon y + iy\log^d y\}J_{\frac{1}{2}(k-2)+\ell+\nu}(2\pi Ry)\,dy.$$

$\psi^{(p)}(t) = \frac{d^p}{dt^p}\psi(t)$. The sum is over non-negative integral values of m, p, j, q, r, and λ such that $m + p + j + q + r + \lambda = \nu$ with the condition that $r < \nu$.

REMARK. The important thing to notice is the following: In each term of the sum for $\mathscr{E}_\varepsilon(t)$ either $p > 0$, or $j + q + \lambda + m > 0$. The proof of Lemma 16 is substantially the same as the proof of Lemma 4. We just remark that the $\mathcal{M}_\varepsilon(t)$ arises from those terms in which all ν derivatives have been applied to the expression $\exp\{it \log^d t\}$. The fact that the two expressions for $\mathcal{M}_\varepsilon(t)$ are the same follows by expanding $(\log^d y + \log^{d-1} y)^\nu$ by the binomial Theorem.

In Theorem 12 below, we shall indicate how $F(x)$ behaves for large $|x|$. We shall write $F(x)$ as a sum of two functions $M_1(x)$ and $M_2(x)$. If $\frac{1}{2}k < b < \frac{1}{2}k + 1$, $M_1(x)$ will be much smaller than $M_2(x)$ for large $|x|$. However the derivatives of $M_2(x)$ will not behave as badly as the derivatives of $M_1(x)$. We shall use these considerations when we study the Lipschitz behavior of the trigonometric series examined in this part of our paper.

THEOREM 12. Let $F(x)$ and $F_\varepsilon(x)$ be the $F(x)$ and $F_\varepsilon(x)$ of Theorem 11. Take ν to be a non-negative integer. Then we can write

3.4 $$F_\varepsilon(x) = M^\nu_{\varepsilon,1}(x) + M^\nu_{\varepsilon,2}(x),$$

where $M^\nu_{\varepsilon,1}$ and $M^\nu_{\varepsilon,2}$ satisfy the following conditions: For large $|x| = R$, $M_1^{(\nu)}(x) = \lim_{\varepsilon \to 0+} M^\nu_{\varepsilon,1}(x)$ exists in the pointwise sense.

3.5 $$M_1^\nu(x) = R^{-\nu-\frac{1}{2}(k-1)} t^{-b+\frac{1}{2}k} \log^{\nu d-c} t$$

$$\cdot\, (\eta_1 \log^{d-1} t + \eta_2 \log^{d-2} t)^{-\frac{1}{2}} Y_\ell(x') \exp\{-idt \log^{d-1} t\}$$

$$\cdot \sum_{j=0}^{\nu} \alpha_j \log^{-j} t(1 + O(t^{-1/6}))$$

as $R \to \infty$. η_1 and α_0 are not zero. t is the largest solution of the equation

3.6 $$2\pi R = \log^d t + d \log^{d-1} t.$$

Finally

3.7 $$|M^\nu_{\varepsilon,1}(x)| = O\{R^{-\nu-\frac{1}{2}(k-1)} t^{-b+\frac{1}{2}k} \log^{\{(\nu-\frac{1}{2})d - c + \frac{1}{2}\}} t\}$$

as $R \to \infty$.

Also for large $|x|$, $M_2^\nu(x) = \lim_{\varepsilon\to 0+} M^\nu_{\varepsilon,2}(x)$ exists, and

3.8 $$|M^{\nu}_{\varepsilon,2}(x)| = O\{R^{-\nu-\frac{1}{2}(k-1)}\}$$

as $R \to \infty$.

Finally if $P = (P_1, \ldots, P_k)$ with P_j a non-negative integer for $j = 1, 2, \ldots, k$, then for sufficiently large ν

3.9 $$|D^P M_2^{\nu}(x)| = O\{R^{-\nu-\frac{1}{2}(k-1)}\}$$

as $R \to \infty$.

REMARK 1. Clearly there exists a y_0 such that $\log^d y + d \log^{d-1} y$ is strictly increasing for $y \geq y_0$. Let $2\pi R_0 = \log^d y_0 + d \log^{d-1} y_0$. We assume $R > R_0$. Then 3.6 has exactly one solution larger than y_0. Also for sufficiently large R,

$$A_d \exp\{(2\pi R)^{1/d}\} \leq t \leq \exp\{(2\pi R)^{1/d}\}.$$

(A_d is some positive constant depending on d.)

REMARK 2. In view of Remark 1 and equation 3.7,

3.10 $$|M^{\nu}_{\varepsilon,1}| \leq A t^{-b+\frac{1}{2}k} R^{-(2c-1)/(2d) + \frac{1}{2}k}.$$

Hence if $b \geq \frac{1}{2}k$ and if $c > \frac{1}{2}$, $|M^{\nu}_{\varepsilon,1}(x)|$ will tend to zero as $x \to \infty$ faster than the reciprocal of any power of R provided d is sufficiently small.

Proof of Theorem 12. We start from equation 3.1, and we apply Lemma 16 to the integral on the right hand side of equation 3.1. We then see that $F_{\varepsilon}(x)$ is a finite linear combination of terms corresponding to the terms of $\mathscr{M}_{\varepsilon}(t)$ and $\mathscr{E}_{\varepsilon}(t)$. We shall only discuss the term which gives the largest contribution to $F_{\varepsilon}(x)$. (The other terms may be examined in a manner either similar to this or trivially.) We shall denote this worst term by $T_{\varepsilon,\nu}(x)$.

3.11 $$T_{\varepsilon,\nu}(x) = 2\pi(-i)^{\ell} Y_{\ell}(x')R^{-\nu-\frac{1}{2}(k-2)} \int_0^{\infty} \psi(y-1) \cdot y^{-b+\frac{1}{2}k}(\log y)^{\nu d-c} \exp\{-\varepsilon y + iy \log^d y\}\, J_{\frac{1}{2}(k-2)+\ell+\nu}(2\pi R y)dy.$$

We shall modify $T_{\varepsilon,\nu}$ to a point where we can use ideas related to Lemmas 13 and 14 (though we cannot use Lemmas 13 and 14 themselves) to prove Theorem 12. At each stage of modification we shall introduce a term $E_{\varepsilon,\nu,j}$. $M^{\nu}_{\varepsilon,2}$ will just be the sum of these "error" terms. We first write

$$T_{\varepsilon,\nu}(x) = G_{\varepsilon,\nu}(x) + E_{\varepsilon,\nu,1}(x).$$

$$E_{\varepsilon,\nu,1}(x) = 2\pi(-i)^{\ell} Y_{\ell}(x')R^{-\nu-\frac{1}{2}(k-2)}$$

$$\cdot \int_0^e \psi(y-1)y^{-b+\frac{1}{2}k}(\log y)^{\nu d-c}\exp(-\varepsilon y + i y \log^d y)J_{\frac{1}{2}(k-2)+\ell+\nu}(2\pi Ry)dy,$$

and

3.13 $$G_{\varepsilon,\nu}(x) = 2\pi(-i)^{\ell} Y_{\ell}(x')R^{-\nu-\frac{1}{2}(k-2)}\int_e^{\infty} y^{-b+\frac{1}{2}k}(\log y)^{\nu d-c}$$

$$\cdot \exp\{-\varepsilon y + i y \log^d y\}J_{\frac{1}{2}(k-2)+\ell+\nu}(2\pi Ry)dy.$$

By Lemma 5

$$|\psi(y-1)J_{\mu}(2\pi Ry)| = O(R^{-\frac{1}{2}}).$$

Therefore it is easy to see

$$E_{\nu,1}(x) = \lim_{\varepsilon\to 0+} E_{\varepsilon,\nu,1}(x) \text{ exists for } x \neq 0, \text{ and that}$$

3.14 $$|E_{\varepsilon,\nu,1}(x)| = O\{R^{-\nu-\frac{1}{2}(k-1)}\}$$

uniformly in $\varepsilon \geq 0$ as $R \to \infty$. If we use again the above inequality for $\psi(y-1)J_{\mu}(2\pi Ry)$, together with the relation

$$\frac{d}{dR}\{J_{\mu}(2\pi Ry)\} = \pi y\{-J_{\mu+1}(2\pi Ry) + J_{\mu-1}(2\pi Ry)\}$$

(see [24], p. 45, equation 2) we see that we can formally differentiate under the integral sign at least ν times and obtain

3.15 $$D^P E_{\nu,1}(x) = O\{R^{-\nu-\frac{1}{2}(k-1)}\}$$

if $\|P\| \leq \nu$.

By Lemma 12, we have

3.16 $$G_{\varepsilon,\nu}(x) = H_{\varepsilon,\nu}(x) + E_{\varepsilon,\nu,2}(x).$$

3.17 $$H_{\varepsilon,\nu}(x) = \int_e^\infty g_1(y,x)\exp\{-\varepsilon y + h(y,R)\}dy,$$

and

$$E_{\varepsilon,\nu,2}(x) = Y_\ell(x')R^{-\nu-\frac{1}{2}+\frac{1}{2}(2-k)}$$

$$\cdot \int_e^\infty g_2(y)\exp(-\varepsilon y + 2\pi iyR)S_2^{\frac{1}{2}(k-2)+\ell+\nu}(2\pi yR)\,dy.$$

In 3.17 and 3.18 we have used the following abbreviations:

3.18 $$h(y,R) = iy\log^d y - 2\pi iyR.$$

3.19 $$g_1(y,x) = 2(-i)^\ell Y_\ell(x')y^{-b+\frac{1}{2}(k-1)}R^{-\nu-\frac{1}{2}(k-1)}$$

$$\cdot (\log y)^{\nu d-c} S_1^{\frac{1}{2}(k-\nu)+\nu+\ell}(2\pi yR).$$

$$g_2(y) = 2(-i)^\ell y^{-b+\frac{1}{2}(k-1)}(\log y)^{\nu d-c}\exp\{iy\log^d y\}.$$

The S's are the S's of Lemma 12.

We shall deal first with the easier integral, $E_{\varepsilon,\nu,2}(x)$. We write for the real variable y a complex variable s, and we change the contour of integration from an infinite ray of the real axis to the contour used in the proof of Theorem 11. One justifies the change of contour in the same way that one did in the proof of Theorem 11. (Except here it is easier because we have essentially replaced $J_{\frac{1}{2}(k-2)+\ell}(2\pi Rs)$ by $\exp(2\pi iRs)$.)

On Λ

$$|g_2(s)| = O\{\exp\{-\tfrac{1}{2}|s|\log^d|s|\}\},$$

as $|s| \to \infty$, since $\arg s = \pi/4$ for $|s| \geq e$ on Λ. Also

$$|\exp(2\pi i s R)| = \exp(-2^{\frac{1}{2}}\pi R|s|)$$

for $|s| \geq e$ and s on Λ. In view of these considerations, we see easily that $E_{\nu,2}(x) = \lim_{\varepsilon \to 0+} E_{\varepsilon,\nu,2}(x)$ exists, and that

3.20 $$|E_{\varepsilon,2,\nu}(x)| = O\{R^{-\nu-\frac{1}{2}(k-1)}\}$$

as $R \to \infty$. Finally by using Lemma 15 and dominated convergence, we can show that

3.21 $$D^P E_{2,\nu}(x) = O\{R^{-\nu-\frac{1}{2}(k-1)}\}.$$

We turn now to $H_{\varepsilon,\nu}(x)$. First we make a change of variables $s = t^{-1}y$ in the integral in 3.17. (t is the t of equation 3.6 in the conclusion of Theorem 12.) We find,

3.22 $$H_{\varepsilon,\nu}(x) = t\int_{et^{-1}}^{\infty} g_1(st, x)\exp\{-\varepsilon\, st + h(st, R)\}\,ds.$$

We shall now think of s as a complex variable. We shall write $s = \sigma + i\tau$ and $s = \rho\exp(i\phi)$. To further investigate $H_{\varepsilon,\nu}(x)$, we shall change the contour of integration from the ray $[et^{-1}, \infty)$ along the positive real axis to a contour Λ_1. We shall describe Λ_1 in terms of an angle θ, $0 < \theta < \pi/4$. We shall be more specific about θ later. The contour Λ_1 consists of four pieces: (See Figure 2). I is the arc of the circle $\rho = et^{-1}$ for which $-\theta \leq \phi \leq 0$. (Remember $s = \rho\exp(i\phi) = \sigma + i\tau$). II is the segment of the ray $\phi = -\theta$ defined by

$$et^{-1} \leq \rho \leq 2^{-\frac{1}{2}}\csc(\pi/4 + \theta).$$

III is the segment of the line $\sigma = \tau + 1$ given by $-\theta \leq \phi \leq \theta$. III will meet II at the point with polar coordinates $(2^{-\frac{1}{2}}\csc(\pi/4 + \theta), -\theta)$. IV is a ray for which $\phi = \theta$, and

$$\rho \geq 2^{-\frac{1}{2}}\csc(\pi/4 + \theta).$$

III and IV meet at the point with polar coordinates $\{2^{-\frac{1}{2}}\csc(\pi/4 - \theta), \theta\}$. The

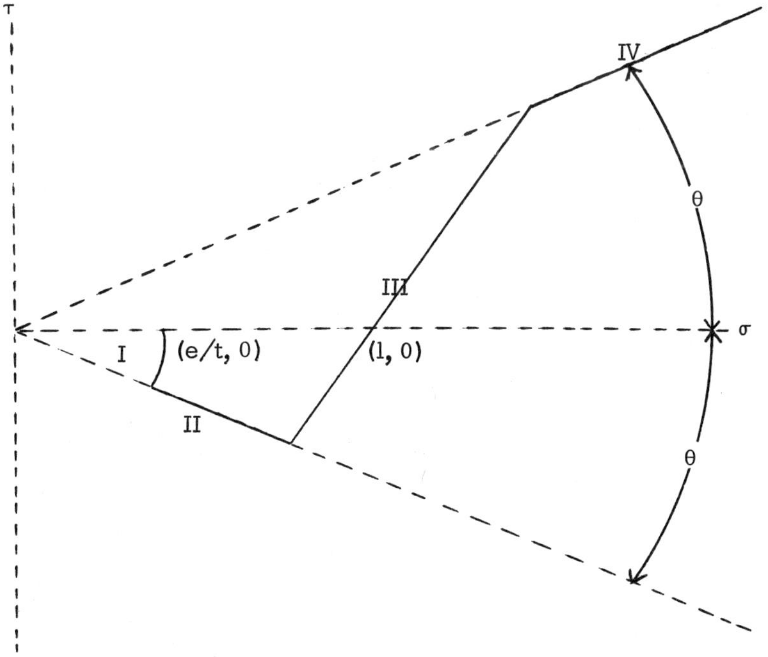

FIGURE 2

main contribution to $M_1^{\nu}(x)$ will come from the portion of the contour labelled III.

We now wish to observe that Cauchy's Theorem may be used to make the desired change in the contour of integration. First, $\mathcal{R}\ell \log s \geq 1$ if $|s| \geq e$. Hence we can find a simply connected region containing c_2, containing the portion of the real axis $\sigma \geq et^{-1}$, and containing the smallest region between them, in which the integrand in 3.22 is single valued. One can also see that for fixed R and ε, the integrand in 3.2 is $O\{\exp(-\varepsilon\sigma t/2)\}$ as $|s| \to \infty$, provided $0 \leq \arg s \leq 3\pi/8$. In view of these considerations it is easy to apply Cauchy's Theorem.

We choose θ, $0 < \theta < \frac{\pi}{4}$, so small that the following conditions should be satisfied:

3.23 $$|1 - s| < 1$$

for s on III.

3.24 $$|s \log s - s - \{-1 + \tfrac{1}{2}(s-1)^2 - (s-1)^3/6\}| < (|s-1|^3)/12$$

for s on III. We can achieve 3.24 because the Taylor series expansion of $s \log s - s$ gives

$$s \log s - s = -1 + \tfrac{1}{2}(s-1)^2 - (s-1)^3/6 + O\{|s-1|^4\}.$$

We shall now consider $h(ts, R)$, defined in equation 3.18. Recalling the definition of t (equation 3.6), we see that

3.25 $$h(st, R) = its\{\log^d(ts) - \log^d t - d\log^{d-1} t\}.$$

Also then,

$$h(st, R) = it\{s(\log t + \log s)^d - s\log^d t - sd\log^{d-1} t\}.$$

If $|\log s/\log t| < \frac{1}{2}$, we see

3.26 $$h(st, R) = it\{sd\log^{d-1} t(\log s - 1)$$
$$+ \tfrac{1}{2}\, sd(d-1)\log^{d-2} t\log^2 s + O\{|s|\log^{d-3} t\log^3 s|\}\}$$

as $t \to \infty$.

We are now in a position to investigate the contribution to $H_{\varepsilon,\nu}$ from the piece of the contour labelled III. We shall denote this contribution by $H_{\varepsilon,\nu,III}$.

($H_{\varepsilon,\nu,III}$ will provide the major contribution to $H_{\varepsilon,\nu}$).

$$H_{\varepsilon,\nu,III}(x) = t \int_{III} g_1(st,x) \exp\{-\varepsilon\ st + h(st,R)\}ds.$$

Clearly $\lim_{\varepsilon\to 0+} H_{\varepsilon,\nu,III} = H_{\nu,III}$ exists, and

$$H_{\nu,III}(x) = t \int_{III} g_1(st,x) \exp\{h(st,R)\}ds.$$

$g_1(st,x)$ (see equation 3.19) is continuous in t and x for x large and s in III. Therefore one can easily see that $H_{\nu,III}(x)$ is continuous for large x.

To further examine $H_{\nu,III}$ we parametrize III by a real variable w. So III is defined by

$$1 + w \exp(i\pi/4)$$

for a certain range of w. Namely

$$-\sin\theta \csc(\pi/4+\theta) \le w \le \sin\theta \csc(\pi/4-\theta).$$

We shall divide III into two parts, III_A and III_B.

$$III_A = \{s \text{ in III} \mid |w| \le t^{-(2/5)}\}.$$

$$III_B = \{s \text{ in III} \mid |w| \ge t^{-(2/5)}\}.$$

Using 3.23, we now observe that for s on III_A,

$$h(st,R) = its\{d \log^{d-1} t(\log s - 1) + \tfrac{1}{2}d(d-1)\log^{d-2} t \log^2 s\} + O(t^{-(1/5)}\log^{d-3} t),$$

as $t\to\infty$. Therefore on III_A,

3.27 $$\exp\{h(st,R)\} = \exp\{itd \log^{d-1} t(s\log s - s)\}$$

$$\cdot \exp\{\tfrac{1}{2}\ its\, d(d-1)\log^{d-2} t \log^2 s\}\{1 + O(t^{-1/5}\log^{d-3} t)\}.$$

Now

$$s\log s - s = -1 + \tfrac{1}{2}(s-1)^2 + O\{|s-1|^3\};$$

hence on III_A,

$$s\log s - s = -1 + \tfrac{1}{2}iw^2 + O(t^{-6/5}).$$

Therefore,

3.28 $$\exp\{itd \log^{d-1} t(s \log s - s)\}$$

$$= \exp\{-itd \log^{d-1} t\}\exp\{-\tfrac{1}{2}td(\log t)^{d-1}w^2\} + O\{t^{-(1/5)}\log^{d-1} t\}.$$

Also

$$s \log^2 s = (1-s)^2 + O(|1-s|^3).$$

So on III_A,

$$s \log^2 s = iw^2 + O(t^{-6/5}).$$

Therefore,

3.29 $$\exp\{\tfrac{1}{2}its\, d(d-1)(\log t)^{d-2} \log^2 s\}$$

$$= \exp\{-\tfrac{1}{2}\, td(d-1)(\log t)^{d-2}w^2\} + O\{t^{-1/5} \log^{d-2} t\}.$$

Substituting 3.28 and 3.29 into 3.27, we observe,

$$\exp\{h(st, R)\} = \exp\{-itd \log^{d-1} t\}$$

$$\cdot \exp\{-w^2(\tfrac{1}{2}td \log^{d-1} t + \tfrac{1}{2}td(d-1)\log^{d-2} t)\} + O(t^{-1/5} \log^{d-1} t).$$

Recalling equation 3.19 one can also see that on III_A

$$g_1(st, x) = 2(-i)^{\ell} R^{-\nu-\frac{1}{2}(k-1)} t^{-b+\frac{1}{2}(k-1)}$$

$$\cdot \log^{\nu d-c} t\, Y_{\ell}(x') \cdot \{\beta^{1}_{\frac{1}{2}(k-2)+\nu+\ell,\, 0} + O(t^{-1/5})\}.$$

(Actually the 1/5 in the O term may be replaced by any number smaller than 2/5.) $\beta^{1}_{\frac{1}{2}(k-2)+\nu+\ell,\, 0}$ is a non-zero constant defined in Lemma 12. Therefore,

$$t \int_{III_A} g_1(st, x)\exp h(st, R)\, ds$$

$$= 2\beta^{1}_{\frac{1}{2}(k-2)+\nu+\ell,\, 0} (-i)^{\ell} Y_{\ell}(x')R^{-\nu-\frac{1}{2}(k-1)} t^{-b+\frac{1}{2}(k+1)}(\log t)^{\nu d-c}$$

$$\cdot \exp\{-itd \log^{d-1} t\} \left\{\int_{-t^{-2/5}}^{t^{-2/5}} \exp\{-w^2 t\Lambda(t)\}dw + O\{t^{-3/5}\log^{d-1} t\}\right\},$$

where

3.30 $$\Lambda(t) = \tfrac{1}{2}d \log^{d-1} t + \tfrac{1}{2}d(d-1)\log^{d-2} t.$$

We now make a change of variables $w' = w(t\Lambda(t))^{\frac{1}{2}}$. Then we easily see that,

3.31 $$t \int_{III_A} g_1(st, x) \exp\{h(st, R)\}ds$$

$$= 2(-i)^{\ell} KR^{-\nu-\frac{1}{2}(k-1)} t^{-b+\frac{1}{2}k} (\log t)^{\nu d-c} (\Lambda(t))^{-\frac{1}{2}} \exp\{-itd \log^{d-1} t\}$$

$$\cdot \{1 + O(t^{-\frac{1}{2}})\}.$$

(Actually the $O(t^{-\frac{1}{2}})$ could be replaced by $O\{t^{-3/5} \log^{d-1} t\}$).

We now consider the contribution from III_B. On III_B $(1-s)^2 = iw^2$. 3.23 implies $|w| \leq 1$ on III. So 3.24 implies

3.32 $$\mathscr{R}\ell \{i(s \log s - s)\} \leq - w^2/4$$

for s on III_B. Also on III

$$|\mathscr{R}\ell\ s \log^2 s| < Aw^2.$$

(Here and in the following discussion A will be a positive number not depending on t or w. A need not be the same in each of its occurrences.) Finally

$$|s \log^3 s| < A|w|^3.$$

It follows now from 3.26 that for s on II_B and t sufficiently large

$$\mathscr{R}\ell\{h(st, R)\} \leq - (d/8)t(\log^{d-1} t)w^2$$

where we have also used the fact that $|w| \geq t^{-2/5}$ for s in III_B. Using again the fact that $|w| \geq t^{-(2/5)}$ on III_B, we see that

$$|\exp\{h(st, R)\}| \leq A \exp\{-At^{1/5}(\log t)^{-1}\}$$

for s on III_B and for t sufficiently large. Furthermore, one can easily see that for s in III and t sufficiently large, $g_1(st, x)$ is uniformly bounded by a power of t. To be more precise

$$|g_1(st, x)| \leq At^{-b+\frac{1}{2}(k-1)}(\log t)^{\nu d-c}$$

for s in III and t sufficiently large. Thus we can easily see

$$\left|t \int_{III_B} g_1(st, x) \exp\{h(st, R)\} ds\right| \leq A \exp(-At(\log t)^{-1})$$

for large t. Recalling equations 3.30 and 3.31, we find

3.33 $$H_{\nu, III}(x) = \alpha_0 R^{-\nu-\frac{1}{2}(k-1)} t^{-b+\frac{1}{2}k} (\log t)^{\nu d-c}$$

$$\cdot\ (\eta_1 \log^{d-1} t + \eta_2 \log^{d-2} t)^{-\frac{1}{2}} Y_\ell(x') \exp\{-idt(\log t)^{d-1}\}\{1 + O(t^{-\frac{1}{2}})\}.$$

$$\alpha_0 = 2(-i)^\ell K = 2(-i)^\ell \beta_{\frac{1}{2}(k-2)+\nu+\ell,\, 0} \int_{-\infty}^{\infty} \exp(-w^2)\, dw.$$

$\eta_1 = \frac{1}{2}d > 0$, and $\eta_2 = \frac{1}{2}d(d-1)$. A similar but simpler argument shows

3.34 $$|H_{\varepsilon, \nu, III}(x)| \leq AR^{-\nu-\frac{1}{2}(k-1)} t^{-b+\frac{1}{2}k} (\log t)^{(\nu-\frac{1}{2})d - c + \frac{1}{2}}.$$

We now discuss the contribution to $H_{\varepsilon, \nu}$ from the portion of Λ_1 labelled IV., i.e., we shall now investigate

$$H_{\varepsilon, \nu, IV}(x) = t \int_{IV} g_1(st, x) \exp\{-\varepsilon st + h(st, R)\} ds.$$

For $\rho = |s|$ large and s on IV,

$$|\log^d(ts) - \log^d(t\rho)| \leq A \log^{d-1}(t\rho).$$

Therefore, using 3.25 we see that for s on IV

$$\mathcal{R}\ell\ \{h(st, R)\} \leq -At\rho \log^d(t\rho)$$

for s sufficiently large, say $\rho > A$. Furthermore

$$|g_1(st, x)| \leq AR^{-\nu-\frac{1}{2}(k-1)} (\rho t)^{-b+\frac{1}{2}(k-1)} (\log(\rho t))^{\nu d-c}.$$

It follows easily therefore that for $R > 0$, $H_{\nu, IV}(x) = \lim_{\varepsilon \to 0+} H_{\varepsilon, \nu, IV}(x)$ exists. It is also clear that

3.35 $$t \left| \int_{\substack{s \text{ in IV} \\ \rho \geq A}} g_1(st, x) \exp(-\varepsilon st + h(st, R)) ds \right| \leq A \exp\{-At\rho \log^d(t\rho)\}.$$

For s in IV with $\rho \geq A$, we use 3.26 to obtain

3.36 $$h(st, R) = itd(\log t)^{d-1}(s - s \log s) + \tfrac{1}{2} s\, d(d-1)(\log t)^{d-2} \log^2 s + O\{|s(\log t)^{d-3} \log^3 s|\},$$

for sufficiently large t.

Suppose for a moment that we could show

$$\mathcal{R}\ell\{i(s \log s - s)\} < 0$$

on that part of IV for which $\rho \leq A$. This is a compact set. Hence,

$$\mathcal{R}\ell\{i(s \log s - s)\} < -A,$$

for some $A > 0$. Hence by 3.36 we would have that for sufficiently large t

$$\mathcal{R}\ell\{h(st, R)\} \leq -At \log^{d-1} t.$$

Also

$$|g_1(st, x)| \leq AR^{-\nu - \frac{1}{2}(k-1)} t^{-b + \frac{1}{2}(k-1)} (\log t)^{\nu d - c}.$$

Hence we can easily demonstrate

3.37 $$\left| t \int_{\substack{s \text{ in IV} \\ \rho \leq A}} g_1(st, x) \exp(-\varepsilon s + h_1(st, R)) ds \right| < A \exp\{-At(\log t)^{d-1}\},$$

for t sufficiently large. Combining this with 3.35, we see that

3.38 $$|H_{\varepsilon, \nu, IV}(x)| < \exp\{-At(\log t)^{d-1}\}$$

for t sufficiently large.

It remains to discuss the sign of $\mathcal{R}\ell\{i(s \log s - s)\}$ on IV. Now

$$\mathcal{R}\ell\{i(s \log s - s)\} = \rho(\sin\phi - \phi \cos\phi - \log \rho \sin\phi).$$

Since $\log\rho$ and $\sin\phi$ are positive on IV, $\mathcal{R}\ell\{i(s \log s - s)\}$ is a decreasing function of ρ for s in IV. Hence we need just check that $\mathcal{R}\ell\{i(s \log s - s)\}$ is negative at the endpoint of IV. But this point is a point of III with $w \neq 0$. We have already mentioned that on III, $\mathcal{R}\ell\{i(s \log s - s)\} < -w^2/4$. (See equation 3.32.) In particular $\mathcal{R}\ell\{i(s \log s - s)\}$ is negative at the point in question.

We turn now to the contribution to $H^{\nu}_{\varepsilon}(x)$ from the other parts of Λ_1. We begin by dividing II into three parts

$$II_A = \{s \text{ in II} \mid \rho \geq \exp(-\theta \cot \theta)\},$$

$$II_B = \{s \text{ in II} \mid (\log t)^{-1} \exp(-\theta \cot\theta) \leq \rho \leq \exp(-\theta \cot\theta)\},$$

and

$$II_D = \{s \text{ in II} \mid et^{-1} \leq \rho \leq (\log t)^{-1} \exp(-\theta \cot\theta)\}.$$

We assume t so large that II_A, II_B, and II_D are all nonvoid. To see the significance of $\exp(-\theta \cot\theta)$, let us denote by $h_1(\rho)$ the real part of $i(s \log s - s)$ with $\phi = -\theta$. (I.e., essentially for s on II.) We observe

3.38 $$h_1(\rho) = \rho \log\rho \sin\theta + \theta \rho \cot\theta - \rho \sin \theta.$$

Now by straightforward differentiation we see the minimum of $h_1(\rho)$ on II is taken at $\rho = \exp\{-\theta \cot\theta\}$. Let

$$H_{\varepsilon, \nu, II_A}(x) = t \int_{II_A} g_1(st, x) \exp\{-\varepsilon st + h(st, R)\} ds.$$

Then

$$H_{\nu, II_A}(x) = \lim_{\varepsilon \to 0+} H_{\varepsilon, \nu, II_A} \text{ exists and}$$

3.39 $$|H_{\varepsilon, \nu, II_A}(x)| \leq A \exp(-At \log^{d-1} t).$$

The reasoning to establish 3.39 is similar to the proof of 3.37, and we shall not repeat the argument.

We turn now to

$$H_{\varepsilon, \nu, II_B}(x) = t \int_{II_B} g_1(st, x) \exp\{-\varepsilon st + h(st, R)\} ds.$$

Since the contour II_B is bounded, it is clear that $H_{\nu, II_B} = \lim_{\varepsilon \to 0+} H_{\varepsilon, \nu, II_B}$ exists. On II_B, $|\log s/\log t| < \frac{1}{2}$. Therefore we may apply 3.26 and observe

$$h(st, R) = idt(\log t)^{d-1}(s \log s - s) + O\{t|s|(\log t)^{d-2} \log^2(\log t)\}$$

for s in II_B. One can now easily check by straightforward differentiation in 3.38 that $h_1(\rho)$ ($=\mathscr{R}\ell\{i(s \log s - s)\}$ for s in II) is a decreasing function of ρ on II_B. Therefore for s in II_B and t sufficiently large,

$$\mathscr{R}\ell\{h(st, R)\} \leq - A \exp\{-At(\log t)^{d-2}(\log \log t)\}.$$

Using again the fact that $g_1(st, x)$ behaves no worse than a power of t, uniformly for s in a compact set, we see

3.40 $$|H_{\varepsilon, \nu, II_B}(x)| < A \exp\{-At(\log t)^{d-2}(\log \log t)\}.$$

We shall now examine

$$E_{\varepsilon, 3}(x) = t \int_{I+II_D} g_1(st, x)\exp\{-\varepsilon\ st + h(st, R)\}ds.$$

It is easy to see that $E_3(x) = \lim_{\varepsilon \to 0+} E_{\varepsilon, 3}(x)$ exists. We shall now estimate the magnitude of $|E_{\varepsilon, 3}|$. In the definition of $E_{\varepsilon, 3}$ we make a change of variables $z = st$ to obtain

$$E_{\varepsilon, 3}(x) = \int_{I' + II'} g_1(z, x)\exp\{-\varepsilon z + h(z, R)\}dz.$$

$$I' = \{z \mid |z| = e \text{ and } -\theta \leq \arg z \leq 0\},$$

and

$$II' = \{z \mid e \leq |z| \leq t(\log t)^{-1} \exp(-\theta \cot\theta) \text{ and } \arg z = \theta\}.$$

Suppose for a moment that we could show

3.41 $$|\exp\{iz \log^d z - 2\pi izR\}| \leq A \exp(-A|z|^{1/2}).$$

We would then have

$$|E_{\varepsilon, 3}(x)| = O\left\{\int_{II'} |g_1(z, x)|\exp\{-A|z|^{1/2}\}|dz|\right\} + O\left\{\max_{z \text{ in } I'} |g_1(z, x)|\right\}$$

by 3.18. In view of 3.19, one then easily sees that

3.42 $$|E_{\varepsilon, 3}(x)| = O\{R^{-\nu-\frac{1}{2}(k-1)}\}.$$

By dominated convergence, we see

$$E_3(x) = \int_{I'+II'} g_1(z,x)\exp\{h(z,R)\}\,dz$$

$$= 2(-i)^{\ell} Y_{\ell}(x')R^{-\nu-\frac{1}{2}(k-1)} \int_{I'+II'} z^{-b+\frac{1}{2}(k-1)}(\log z)^{\nu d-c}$$

$$\cdot \exp\{iz \log^d z - 2\pi izR\}\, S_1{}^{\frac{1}{2}(k-2)+\nu+\ell}(2\pi zR)\,dz.$$

Now using Lemma 15, 3.41, and dominated convergence, we can justify formal differentiation in the above expression for $E_3(x)$ ν times. We thus see that

3.43
$$|D^P E_3(x)| \le AR^{-\nu-\frac{1}{2}(k-1)}$$

for large R, if $\|P\| \le \nu$.

We shall now prove 3.41. If $|z| \le t^{\frac{1}{2}}$, and z is in II',

$$\mathcal{R}\ell\,\{iz(\log z)^d - 2\pi izR\} = -|z|\sin\theta\,\{2\pi R - (\log|z|)^d + O(\log|z|)^{d-1}\}$$

$$\le -|z|\sin\theta\,\{2\pi R - (\log(\exp(2\pi R)^{1/d})^{1/2})^d$$

$$+ O((\log(\exp(2\pi R)^{1/d})^{1/2})^{d-1})\}$$

$$\le -|z|\sin\theta\,\{2\pi R - (\tfrac{1}{2})^d(2\pi R) + O(2\pi R)^{(d-1)/d}\}$$

$$\le -A|z|\sin\theta,$$

if R is sufficiently large. (In the above argument we used the fact that $t \le \exp\{(2\pi R)^{1/d}\}$; this is discussed in Remark 1 following Theorem 12.)

If $t^{\frac{1}{2}} \le |z| \le t(\log t)^{-1}$, we see, by similar reasoning, that

$$\mathcal{R}\ell\,\{iz(\log z)^d - 2\pi izR\}$$

$$\le -|z|\sin\theta\,\{2\pi R - [(2\pi R)^{1/d} - \log(2\pi R)]^d + O((2\pi R)^{(d-1)/d})\}$$

$$\le -d|z|\sin\theta\,\{(2\pi R)^{(d-1)/d}\log(2\pi R) + O((2\pi R)^{(d-1)/d})\}$$

$$\le -A|z|\sin\theta(2\pi R)^{(d-1)/d}$$

(for R large)

$$\leq - A|z|^{\frac{1}{2}}$$

since $|z| \geq t^{\frac{1}{2}}$.

We have now divided $H^{\nu}_{\varepsilon}(x)$ into a number of pieces. Namely

$$H^{\nu}_{\varepsilon}(x) = H_{\varepsilon,\nu,III}(x) + H_{\varepsilon,\nu,IV}(x) + H_{\varepsilon,\nu,II_A}(x) + H_{\varepsilon,\nu,II_B}(x) + E_{\varepsilon,\nu,3}(x).$$

So from 3.12 and 3.16, we have,

3.44 $$T^{\nu}_{\varepsilon}(x) = H_{\varepsilon,\nu,III}(x) + H_{\varepsilon,\nu,IV}(x) + H_{\varepsilon,\nu,II_A}(x) + H_{\varepsilon,\nu,II_B}(x)$$

$$+ E_{\varepsilon,\nu,1}(x) + E_{\varepsilon,\nu,2}(x) + E_{\varepsilon,\nu,3}(x).$$

We have shown that the limit as $\varepsilon \to 0+$ of the various E's and H's in 3.44 exist, at least for sufficiently large x. We write

$$M^{\nu}_{\varepsilon,2}(x) = E^{\nu}_{\varepsilon,1}(x) + E^{\nu}_{\varepsilon,2}(x) + E^{\nu}_{\varepsilon,3}(x).$$

Then equations 3.14, 3.20 and 3.42 show that 3.8 is satisfied. Furthermore, equations 3.15, 3.21, and 3.43 demonstrate 3.9. We set

$$M^{\nu}_{\varepsilon,1} = H^{\nu}_{\varepsilon,III} + H^{\nu}_{\varepsilon,IV} + H^{\nu}_{\varepsilon,II_A} + H^{\nu}_{\varepsilon,II_B}.$$

Then equation 3.7 is a consequence of 3.34, 3.35, 3.39, and 3.40. Finally equation 3.5 (with $\alpha_0 \neq 0$, $\alpha_1 = \alpha_2 = \ldots \alpha_{\nu} = 0$) follows from 3.33, 3.35, 3.39, and 3.40. When we examine the other terms into which we resolved $F_{\varepsilon}(x)$ by means of Lemma 16, we obtain other terms for which $\alpha_0 = 0$ (in equation 3.5), but for which some other of the α_j may not be zero. Hence we establish Theorem 12.

We shall now use Theorems 11 and 12 to study trigonometric series.

THEOREM 13. Let

$$f_{\varepsilon}(x) = \sum{}' |n|^{-b} \log^{-c}|n| Y_{\ell}(n')\exp\{-\varepsilon|n| + i|n|\log^{d}|n| + 2\pi i n \cdot x\}.$$

Assume either

$$\text{i)}\ \ b > \tfrac{1}{2}k$$

or

$$\text{ii)}\ \ b = \tfrac{1}{2}k$$

and

$$(2c-1) > dk.$$

Then $f(x) = \lim_{\varepsilon \to 0+} f_\varepsilon(x)$ exists in the pointwise sense, and $f(x)$ is continuous. Furthermore if i) or ii) is satisfied

$$f(x) \sim \sum{}' |n|^{-b} \log^{-c}|n| Y_\ell(n')\exp\{i|n|\log^d|n| + 2\pi i n \cdot x\}.$$

Theorem 13 is an immediate consequence of Theorem 11, Theorem 12, and Lemma 11 of part I. The only point that needs to be verified is that for some $\mu > 0$, $|F_\varepsilon(x)| = O(|x|^{-k-\mu})$ uniformly in ε as $x \to \infty$. This point follows from equations 3.8, 3.9 and Remark 1 to Theorem 12 if $\nu > \frac{1}{2}k$.

Theorem 14 will demonstrate various smoothness properties of the $f(x)$ of Theorem 13.

THEOREM 14. Let $f(x)$ be the $f(x)$ of Theorem 13.

i) Let r be a positive integer. Assume either

a. $b > \frac{1}{2}k + r$

or

b. $b = \frac{1}{2}k + r$

and

$$2c-1 > dk$$

Then the partial derivatives of $f(x)$ of order r exist and are continuous. In particular, b) above can be satisfied for some $d > 0$ if $c > \frac{1}{2}$ and if $b = \frac{1}{2}k + r$.

ii) Let $0 < a < 1$. Suppose $b = \frac{1}{2}k + a$ and $d(k-2) + 1 \leq 2c$. Then $f(x)$ is in Lip a. In particular if $c > \frac{1}{2}$, and if $b = \frac{1}{2}k + a$, there exists a $d > 0$ such that f is in Lip a.

To prove Theorem 14, we need a lemma enabling us to calculate the derivatives of f(x) and F(x). (F(x) is the F(x) of Theorems 11 and 12).

LEMMA 17. Let F(x) be as in Theorems 11 and 12, and let f(x) be as in Theorem 13.

i) F(x) is infinitely differentiable, and any r^{th} order derivative of F(x) is of the form

$$\sum_j \beta_j \lim_{\varepsilon \to 0+} \left\{ \psi(|x| - 1)|x|^{r-b} \log^{-c}|x| Y_\ell(x') \exp\{i|x|\log^d|x|\} \right\}^{\vee}.$$

The range of summation is finite, and the β_j are some numbers.

ii) If oither

$$\text{a.} \quad b > \tfrac{1}{2}k + r,$$

or if

$$\text{b.} \quad b = \tfrac{1}{2}k + r$$

and

$$2c-1 > dk,$$

then the rth order derivatives of f(x) exist and are of the form

$$\sum_j \beta_j \lim_{\varepsilon \to 0+} \left\{ \sum{}' |n|^{r-b} \log^{-c}|n| \, Y_\ell(n') \exp(-\varepsilon|n| + i|n|\log^d|n| + 2\pi i n \cdot x) \right\}.$$

The range of summation is finite.

Proof of Lemma 17. The product of two spherical harmonics of finite degree is a finite linear combination of spherical harmonics. (This is a consequence of Proposition 3-b on p. 31 of [6].) In particular if $Y_\ell(x')$ is a spherical harmonic $x_i(|x|)^{-1} Y_\ell(x')$ is a finite linear combination of spherical harmonics. Therefore the proof of Lemma 17 is the same as the proof of Theorem 5 of part I.

Proof of Theorem 14. i) follows immediately from Part ii) of Lemma 17 and Theorem 13.

To prove ii) we employ Lemma 11 of part I. The proof of Lemma 11' shows that except for a trigonometric polynomial which we may neglect

$$f(x) = F(x) + E(x)$$

where

$$E(x) = \sum_{n\neq 0} F(x + 2\pi n).$$

Let $h = (h_1, \ldots, h_k)$ be a vector in k-dimensional space, which should be thought of as small. (We shall assume $|h| < 1$.) We may write

$$f(x+h) - f(x) = B(x,h) + C(x,h) + D(x,h) + E(x,h);$$

where

$$B(x,h) = \sum_{|n+x| \leq \xi-3} \{F(x+n+h) - F(x+n)\},$$

$$C(x,h) = \sum_{|n+x| \geq \xi+3} \{M_1^{\nu}(x+n+h) - M_1^{\nu}(x+h)\},$$

$$D(x,h) = \sum_{|n+x| \geq \xi+3} \{M_2^{\nu}(x+n+h) - M_2^{\nu}(x+h)\},$$

and

$$E(x,h) = \sum_{\xi-3 < |n+x| < \xi+3} \{F(x+n+h) - F(x+n)\}.$$

The M_1^{ν} and M_2^{ν} are as in Theorem 1, and we choose $\xi = (2\pi)^{-1}\{\log(1/|h|)\}^d$. In the following discussion, A will denote a suitably large constant, not depending on n, x, or h, which may vary from place to place.

We first look at $B(x, h)$.

$$|F(x+n+h) - F(x+n)| \leq A|h| \max_{1\leq j\leq n} \left.\frac{\partial}{\partial x_j} F(x)\right|_{x=\zeta},$$

where $\zeta = (\zeta_1, \ldots, \zeta_k)$ with $x_j + n_j \leq \zeta_j \leq x_j + n_j + h_j$ (if the $h_j \geq 0$). Hence using Lemma 17 (with r = 1), and Theorem 12 (with any ν) (see Remarks I and II), we see

$$|B(x,h)| \leq A|h| \sum_{|n+x| \leq \xi-3} |n|^{-(2c-1)/2d+\frac{1}{2}k}[\exp\{(2\pi|n+x|)^{1/d}\}]^{-b+1+\frac{1}{2}k},$$

(since $|n+x|$ is less than any power of h^{-1}, $\exp\{(2\pi|\zeta|)^{1/d}\}$ is of the same order of magnitude as $\exp\{(2\pi|x+n|)^{1/d}\}$.) In view of the hypothesis on b, c, and d, we see

$$|B(x,h)| \leq A|h| \sum_{0<|n+x| \leq \xi-3} |n|^{1-k}[\exp\{(2\pi|n+x|)^{1/d}\}]^{1-a}$$

$$\leq A|h| \int_{|y|\leq \xi-1} |y|^{1-k}[\exp\{(2\pi|y|)^{1/d}\}]^{1-a} dy$$

$$\leq A|h| \int_0^{\xi-1} [\exp\{(2\pi u)^{1/d}\}]^{1-a} dy \leq A|h|[\exp\{(2\pi\xi)^{1/d}\}]^{1-a} \leq A|h|\,|h|^{a-1} \leq A|h|^a.$$

For any fixed ν,

$$|C(x,h)| \leq \sum_{|n+x| \geq \xi+3} |M_1^{\nu}(x+n+h)| + |M_1^{\nu}(x+n)|.$$

Using Theorem 12 (in particular Remark I)

$$|C(x,h)| \leq A \sum_{|n+x| \geq \xi+3} |n|^{-(2c-1)/2d+\frac{1}{2}k}[\exp\{(2\pi|n+x|)^{1/d}\}]^{-b+\frac{1}{2}k}$$

since $|M_1^{\nu}(x+n+h)| = O(|M_1^{\nu}(x+h)|)$ if $|n+x|$ is less than an appropriate power of $(1/|h|)$, and the terms with $|n+x|$ larger than any given positive power of $(1/|h|)$ make a negligible contribution. So

$$|C(x,h)| \leq A \sum_{|n|>\xi+3} |n|^{1-k}[\exp(2\pi|n+x|)^{1/d}]^{-a} \leq A[\exp\{(2\pi\xi)^{1/d}\}]^{-a} \leq Ah^a.$$

Since we still have ν at our disposal, it is easy to show

$$|D(x,h)| \leq A|h|,$$

if ν is sufficiently large, by using the Mean Value Theorem and Theorem 12.

The number of terms in $E(x,h)$ is $O(n^{k-1})$. Hence one need only observe

$$|\exp\{(2\pi|n+x+h|)^{1/d}\} - \exp\{(2\pi|n+x|)^{1/d}\}| \leq A|h|^a \quad \text{for } \xi-3 < |n+x| < \xi+3.$$

If both $|n+x+h|$ and $|n+x|$ are less than or equal to ξ, one makes this observation via the Mean Value Theorem. If both $|n+x|$ and $|n+x+h|$ are greater than or equal to ξ, the estimate is obvious. In general, one may write

$$\exp\{(2\pi|n+x+h|)^{1/d}\} - \exp\{(2\pi|n+x|)^{1/d}\}$$

$$= \exp\{(2\pi(|n+x+h|)^{1/d}\} - \exp\{(2\pi|n+x+h_1|)^{1/d}\} + \exp\{(2\pi|n+x+h_1|)^{1/d}\}$$

$$- \exp\{(2\pi|n+x|)^{1/d}\}$$

where $|h_1| \leq Ah$ and $|n+x+h_1| = \xi$.

PART IV

In part IV we shall discuss various applications of parts I, II, and III. Our first applications concern the absolute convergence of Fourier series.

A classical result of S. Bernstein asserts the following: (see [3]).

Let $f(t)$ be a periodic function of a real variable t. Suppose for some $\varepsilon > 0$, $f(t)$ is in Lip $(\frac{1}{2} + \varepsilon)$. Then the Fourier series of f converges absolutely.

Bernstein also showed that there exists a function f, in Lip $\frac{1}{2}$ such that the Fourier series of f does not converge absolutely (see [3]).

The following result is known; see [5].

If $f(x)$ is defined on T^k, and if $f(x)$ is in Lip $\{\frac{1}{2}k + \varepsilon\}$, then the Fourier series of f converges absolutely.

We shall prove

THEOREM 15. There exists a function $f(x)$ in Lip$\{\frac{1}{2}k\}$ such that the Fourier series of f does not converge absolutely.

Szasz and Minakshisundaram [20] have obtained the following result.

Let $0 < a < 1$. There exists an $f(x)$ defined on T^k with Fourier coefficients $c(n)$ such that f is in Lip a and

$$\sum |c(n)|^{2k/(k+2a)} = \infty.$$

We shall treat the case $1 \leq a \leq \frac{1}{2}k$.

THEOREM 16. Let $1 \leq a \leq \frac{1}{2}k$. There exists a function $f(x)$ in Lip a with Fourier coefficients $c(n)$ such that

$$\sum |c(n)|^{2k/(k+2a)} = \infty.$$

REMARK: If we put $a = \frac{1}{2}k$, we see that Theorem 16 contains Theorem 15.

Proof of Theorem 16. Let $f(x)$ have Fourier coefficients $c(n)$ where

$$c(n) = 0$$

if $|n| \leq 3/2$, and

$$c(n) = |n|^{-a-\frac{1}{2}k}(\log |n|)^{-(k+a)/(2k)} \exp\{i|n| \log^{d}|n|\}$$

if $|n| > 3/2$. $d > 0$. (There is a unique f in L^2 with Fourier coefficients $c(n)$ since $\Sigma |c_n|^2 < \infty$.) Now $(k+a)/(2k) > \frac{1}{2}$. Therefore by part i) of Theorem 14 we can choose d so small that the partial derivatives of f of order $[a]$ exist and are continuous. Again $(k+a)/(2k) > \frac{1}{2}$, so by Lemma 17 and by part ii) of Theorem 14, we can choose d so small that the derivatives of f of order $[a]$ are in $\mathrm{Lip}\{a - [a]\}$. Thus f is in Lip a. But clearly

$$\sum |c(n)|^{2k/(k+2a)} = \infty.$$

Our next application is to the theory of multipliers. We shall say that a bounded set of numbers $\{\gamma(n)\}$ (where n runs over the lattice points of E_k) is a multiplier on $L^p(T^k)$ if the following condition is satisfied: Whenever

$$\sum c(n)\exp(2\pi i n \cdot x)$$

is the Fourier series of a function in $L^p(T^k)$, then

$$\sum \gamma(n)c(n)\exp(2\pi i n \cdot x)$$

is the Fourier series of a function in $L^p(T^k)$. Thus a multiplier on L^p induces a natural transformation from $L^p(T^k)$ to $L^p(T^k)$.

Hirschman [14] has shown the following one dimensional result:

Let $d(0) = 0$, and for $n \neq 0$ define

$$d(n) = |n|^{-b}\exp(2\pi i|n|^a)$$

where $0 < a < 1$ and $b < \frac{1}{2}a$. Then $\{d(n)\}$ is a multiplier on $L^p(T^1)$ if

$$2a/(a + 2b) < p < 2a(a - 2b).$$

If

$$p < 2a/(a + 2b),$$

or if $$p > 2a/(a - 2b),$$
then $\{d(n)\}$ is not a multiplier on $L^p(T^1)$.

We generalize Hirschman's result to k-dimensions.

THEOREM 17. Let $d(0) = 0$, and for $n \neq 0$, let
$$d(n) = |n|^{-b} \exp\{2\pi i |n|^a\}.$$
Let $0 < a < 1$, and assume $b < \frac{1}{2}ka$. Then

i) $\{d(n)\}$ is a multiplier on $L^p(T^k)$ if
$$2ka/(ka + 2b) < p < 2ka/(ka - 2b).$$

ii) If either
$$p < 2ka/(ka + 2b),$$
or if
$$p > 2ka/(ka - 2b),$$
then $\{d(n)\}$ is not a multiplier on $L^p(T^k)$.

We shall need to know several facts about multiplier transformations:

4.1. Any bounded set $\{\gamma(n)\}$ is a multiplier on L^2. The L^2 norm of the transformation induced by $\{\gamma(n)\}$ is $\sup_n \{|\gamma(n)|\}$.

4.1 is a well known immediate consequence of Parseval's relation, and the Riesz-Fischer Theorem.

4.2. The set $\{\gamma(n)\}$ is a multiplier on L^p if and only if the set $\{\gamma(n)\}$ is a multiplier on L^q, where $1/p + 1/q = 1$.

4.2 is proved by a well-known duality argument.

4.3. Let $\gamma(0) = 0$, and set
$$\gamma(n) = \exp\{i\pi y \log |n|\}$$
with y real for $n \neq 0$. Then the set $\{\gamma(n)\}$ is a multiplier on L^p for $1 < p < \infty$. Moreover the norm of the transformation defined by $\{\gamma(n)\}$ is equal

to or less than $A(p)(1 + |y|)^k$ for a fixed p.

The proof of 4.3 is an immediate consequence of the Marcinkiewicz multiplier Theorem. (See [17]).

4.4. Stein's interpolation Theorem for an analytic family of operators (see [20]).

We say that a family of operators $\{T_z\}$ is an analytic family of operators if the following four conditions hold.

i) $T_z(f)$ is defined whenever f is a simple function and $0 \leq \mathscr{R}\ell\, z \leq 1$.

ii) For any other simple function g(x),

$$I(z) = \int_{T^k} g(x) T_z(f(x))\, dx$$

exists.

iii) I(z) is an analytic function of z for $0 < \mathscr{R}\ell\, z < 1$.

iv) I(z) is a continuous function of z for $0 \leq \mathscr{R}\ell\, z \leq 1$. An important special case of Stein's Interpolation Theorem follows:

Let $T_z f$ be an analytic family of linear operators. Let $p \geq 1$, and let $r \geq 1$. Suppose

$$\| T_{iy} f \|_p \leq M_1(y) \| f \|_p$$

and

$$\| T_{1+iy} f \|_r \leq M_2(y) \| f \|_r$$

for each simple f, where

$$\log M_j(y) \leq A \exp(ay) \qquad (a < \pi;\ j = 1, 2).$$

Then

$$\| T_t(f) \|_\rho \leq A(t) \| f \|_\rho$$

where

$$1/\rho = t/p + (1-t)/r,$$

and A(t) does not depend on f.

4.5. Let $\{\gamma(n)\}$ be a bounded set of numbers. Let f be a simple function with Fourier coefficients $c(n)$. Define $T_z(f)$ to be the function with n^{th} Fourier coefficient $\gamma(n)c(n)$. Then $T_z(f)$ is well-defined and is an analytic family of operators.

The proof of 4.5 is trivial.

4.6. If the set $\gamma(n)$ is the set of Fourier coefficients of a function $f(x)$ in L^1, the set $\{\gamma(n)\}$ is a multiplier on every L^p. Moreover, the L^p norm of the transformation defined by $\{\gamma_n\}$ is at most the L^1 norm of f.

4.6 is well known. The proof is a consequence of the convolution structure of trigonometric series.

4.7. Let $\{\gamma(n)\}$ be a bounded set. Suppose f is any simple function, and denote its Fourier coefficients by $c(n)$. Let Tf be the function with Fourier coefficients $\gamma(n)c(n)$. Suppose there is a number $A(p)$ such that

$$\|Tf\|_p \leq A(p)\|f\|_p.$$

Then $\{\gamma(n)\}$ is a multiplier on L^p, and the norm of the transformation induced by $\{\gamma(n)\}$ is at most A_p.

4.7 is a well-known consequence of a duality argument, and the fact that a function uniformly continuous on a dense subset of a metric space has a unique continuation to a continuous function on the whole space.

4.8. If a bounded set $\{\gamma(n)\}$ is a multiplier on L^p for $1 \leq p \leq 2$, then $\{\gamma(n)\}$ is a multiplier on L^r for $p \leq r \leq 2$. The proof of 4.8 uses the Banach Steinhaus Theorem, the M. Riesz Interpolation Theorem and 4.7. 4.8 is common folklore.

The following Lemma will essentially contain part i) of Theorem 17.

LEMMA 18. Let $\gamma_0 = \{\gamma(n)\}$ be a bounded set of numbers with $\gamma(0) = 0$. Say $|\gamma(n)| \leq M$. Define γ_β to be the set of numbers $\{\gamma(n)|n|^{-\beta}\}$. Suppose for some fixed μ, and every $\varepsilon > 0$, $\gamma_{\mu+\varepsilon}$ is the set of Fourier coefficients of a function in $L^1(T^k)$. Then if $0 < \rho < \mu$, and if

$$2\mu(\mu + \rho)^{-1} < p < 2\mu(\mu - \rho)^{-1},$$

γ_ρ is a multiplier on L^p.

Proof of Lemma 18. By 4.2, it suffices to prove Lemma 18 for p such that $1 < p \leq 2$. Also 4.1 implies that γ_β is a multiplier on L^2 for any $\beta \geq 0$.

Let $f(x)$ be a simple function with Fourier coefficients $c(n)$. We define $T_z(f)$ to be the function with Fourier coefficients $d(n)$, where

$$d(0) = 0$$

and

$$d(n) = \gamma(n)c(n)|n|^{-(\mu+\varepsilon)z}$$

for $n \neq 0$. (We assume $\varepsilon > 0$, and $0 \leq \mathscr{R}\ell\, z \leq 1$.) By 4.5 $T_z(f)$ is well defined and is an analytic family of operators in the sense of 4.4. The set $\gamma_{(\mu+\varepsilon)iy}$ is bounded by M. Hence by 4.1

4.9 $$\|T_{iy}f\|_2 \leq M\|f\|_2.$$

Let δ be any positive number less than one. Suppose for a moment we could show

4.10 $$\|T_{1+iy}f\|_{1+\delta} \leq A(\delta, \varepsilon)(1 + |y|^k)\|f\|_{1+\delta}.$$

Using 4.9 and 4.10, and applying 4.4 with $p = 1 + 8$, $\delta < 1$, and $r = 2$, we observe

$$\|T_t f\|_\tau \leq A(t, \tau)\|f\|_\tau,$$

$0 \leq t \leq 1$, and

$$\tau = 2(1 + \delta)/\{1 + \delta + (1-\delta)t\}.$$

($T_t f$ is the function with Fourier coefficients $\{\gamma(n)c(n)|n|^{-(\mu+\varepsilon)t}\}$). By 4.7, then, $\gamma_{(\mu+\varepsilon)t}$ is a multiplier on L^τ. Since $\delta < 1$, $\tau < 2$. We thus have by 4.8, that $\gamma_{(\mu+\varepsilon)t}$ is a multiplier on L^p for $2(1 + \delta)\{1 + \delta + (1-\delta)t\}^{-1} < p \leq 2$. Since δ can be arbitrarily small, $\gamma_{(\mu+\varepsilon)t}$ is a multiplier on L^p for $2 \geq p > 2/(1 + t)$. If we now put $\rho = (\mu+\varepsilon)t$, we see γ_ρ is a multiplier on L^p

for $2 \geq p > 2\{1+\rho/(\mu+\varepsilon)\}^{-1}$. Since ε is any positive number, we find that γ_ρ is a multiplier on L^p if $2 \geq p > 2\{1+\rho/\mu\}^{-1} = 2\mu/(\mu+\rho)$.

All we must do now is establish 4.10. We have a simple function $f(x)$ with Fourier coefficients $c(n)$. Let $g(x)$ be the function with Fourier coefficients $c(n)\gamma(n)|n|^{-(\mu+\varepsilon)}\exp(2\pi i y \log|n|)$. We must show

4.11 $$\|g(x)\|_{1+\delta} \leq A(\delta,\varepsilon)(1+|y|^k)\|f(x)\|_{1+\delta}.$$

By hypothesis $\{\gamma_n|n|^{-\mu-\varepsilon}\}$ are the Fourier coefficients of a function $\theta_\varepsilon(x)$ in $L^1(T^k)$. Hence by 4.6, the numbers $\{c(n)\gamma(n)|n|^{-\mu-\varepsilon}\}$ are the Fourier coefficients of a function $Q(x)$ in $L^{1+\delta}$. ($Q(x)$ is even in L^1). Moreover

4.12 $$\|Q(x)\|_{1+\delta} \leq A(\varepsilon)\|f(x)\|_{1+\delta}.$$

But 4.3 shows that 4.12 implies 4.11. This finishes the proof of Lemma 18.

Proof of Theorem 17. Let

$$\gamma(0) = 0,$$

and let

$$\gamma(n) = \exp\{2\pi i|n|^a\}$$

for $n \neq 0$. Let $\varepsilon > 0$. By Theorem 10 of part II, the set

$$\{\gamma(n)|n|^{-\frac{1}{2}ka-\varepsilon}\}$$

is the set of Fourier coefficients of a function in L^1. We now apply Lemma 18 with $\mu = \frac{1}{2}ka$, and we obtain assertion i) of Theorem 17.

We now turn to assertion ii). By 4.2, it suffices to consider the case $1 < p \leq 2$. Let $k > \lambda > 0$. Set

$$h(0) = 0,$$

and let

$$h(n) = |n|^{-\lambda}$$

for $n \neq 0$. By Theorem 7 of part I, there exists a function $H(x)$ in $L^1(T^k)$ with Fourier coefficients $h(n)$. Moreover $H(x)$ is continuous in T^k for $x \neq 0$ and

$$H(x) \simeq A(\lambda)R^{\lambda-k}$$

as $x \to 0$ with $A(\lambda) \neq 0$. So $H(x)$ is in L^p if

4.13 $$p(k-\lambda) < k.$$

Now let $b \geq \frac{1}{2}ak$. Define

$$u(0) = 0,$$

and

$$u(n) = |n|^{-b-\lambda} \exp\{2\pi i |n|^a\}$$

for $n \neq 0$. By Theorem 10 of part II there exists a $U(x)$ in $L^1(T^k)$ with Fourier coefficients $u(n)$. Also

$$U(x) \simeq K(\lambda, R) R^{-(k - \frac{1}{2}ka - b - \lambda)/(1-a)}$$

as $x \to 0$, where $K(\lambda, R)$ is bounded away from zero. So $U(x)$ is not in L^p if

$$p(k - \tfrac{1}{2}k - b - \lambda) = k(1 - a).$$

So $U(x)$ is not in L^p if

$$\lambda = (k/p)(a - 1) - b - \tfrac{1}{2}ak + k.$$

In view, then, of 4.13 $\{d(n)\}$ is not a multiplier on L^p if

$$p\{(1-a)k/p + \tfrac{1}{2}ak + b\} < k;$$

that is if

$$p < ak/\{\tfrac{1}{2}ak + b\} = 2ak/\{ak + 2b\}.$$

This finishes the proof of Theorem 17.

Our next topic will be fractional integration. Let $f(x)$ be in $L^1(E_k)$. Du Plessis [10] gives the following definition of $f_\alpha(x)$, the Riesz fractional integral of $f(x)$ of order $\alpha > 0$, $0 < \alpha < k$:

4.14 $$f_\alpha(x) = C(k, \alpha) \int_{E_k} |x-y|^{\alpha-k} f(y)dy,$$

where

$$C^{-1}(k, \alpha) = \pi^{\frac{1}{2}k} \Gamma(\tfrac{1}{2}\alpha) \{\Gamma(\tfrac{1}{2}(k-\alpha))\}^{-1}.$$

He then proves several theorems about $f_\alpha(x)$. The following three are analogues of classical one-dimensional theorems:

4.15. THEOREM A: If $f(x)$ is in $\mathrm{Lip}\,\beta$, $0 < \beta < 1$, then $f_\alpha(x)$ is in $\mathrm{Lip}\,(\alpha+\beta)$ provided $\alpha + \beta < 1$.

4.16. THEOREM B: If $f(x)$ is in L^q, $q > 1$, and $1 + k/q > \alpha > k/q$, then $f_\alpha(x)$ is in $\mathrm{Lip}(\alpha - k/q)$.

4.17. THEOREM C: If $f(x)$ is in L^q and $0 < \alpha < k/q$, then $f_\alpha(x)$ is in L^r, where $\alpha = k(1/q - 1/r)$. (I think that Du Plessis means $q > 1$ in Theorem C as well as in Theorem B.)

The definition 4.14 is not suitable for periodic functions because, in general, the integral will not exist. Now if we could proceed formally from equation 4.20, we would obtain

4.18 $$f_\alpha\hat{}(x) = |x|^{-\alpha} f\hat{}(x).$$

We shall define fractional integration for periodic functions from the point of view of 4.18. We shall then prove that $f_\alpha(x)$ can be represented by a convolution transformation similar to equation 4.20.

4.19. DEFINITION. Let $f(x)$ be in $L^1(T^k)$. Designate the Fourier coefficients of f by $c(n)$. Let $0 < \alpha < k$. $f_\alpha(x)$ is defined to be the unique function in L^1 with Fourier coefficients $d(n)$, where

$$d(0) = 0,$$

and

$$d(n) = c(n)|n|^{-\alpha}$$

for $n \neq 0$.

We shall prove that the above definition makes sense. First, however, we remark that the one dimensional Weyl fractional integral for periodic functions is given by a multiplier transformation $\{\gamma(n)\}$, where

$$\gamma(0) = 0$$

and

$$\gamma(n) = (in)^{-\alpha}$$

for $n \neq 0$. (See [25], Vol. II, p. 133.)

We shall prove the following theorems about fractional integration:

THEOREM 18. $f_\alpha(x)$ is well defined for $f(x)$ in $L^1(T^k)$ by the definition of 4.19. Also

$$f_\alpha(x) = \int_{T^k} g_\alpha(x-y)\, f(y)\, dy. \tag{4.20}$$

$g_\alpha(x)$ is periodic, and

$$g_\alpha(x) = C(k,\alpha)|x|^{\alpha-k} + E_\alpha(x); \tag{4.21}$$

where $E_\alpha(x)$ is infinitely differentiable for x in T^k.

THEOREM 19. If we define $f_\alpha(x)$ by 4.19, then 4.15 (Theorem A), 4.16 (Theorem B), and 4.17 (Theorem C) hold.

We add a few remarks on Theorems C and D.

THEOREM 20.

i) In Theorem B, the conditions $q > 1$, k/q strictly less than α, and α strictly less than $1 + k/q$ are necessary.

ii) In Theorem C, the conditions $q > 1$ and α strictly less than k/q are necessary.

Theorem 20 is valid both for fractional integration on E_k with the definition 4.14 and in the case of fractional integration in $L^1(T^k)$ with definition 4.19.

REMARK: When, in Theorem 20, we are required to construct an example $g(x)$, possessing certain properties, but such that the derivatives of $g(x)$ of order r are not continuous (i.e., such that $g(x)$ does not belong to Lip r), our proof will actually show a little more. Namely, at least one of the rth derivatives of the example $g(x)$ will not be locally bounded.

<u>Proof of Theorem 18.</u> We define

$$g_\alpha(x) = \lim_{\varepsilon \to 0+} \sum{}' |n|^{-\alpha} \exp\{-\varepsilon n + 2\pi i n \cdot x\}.$$

By the remark to Theorem 7 of part I, equation 4.21 is valid. Moreover Theorem 7 asserts that the Fourier coefficients of $g_\alpha(x)$ are given by

4.22 $$d(n) = \begin{cases} 0 & \text{if } n = 0 \\ |n|^{-\alpha} & \text{if } n \neq 0 . \end{cases}$$

Let $f(x)$ be in $L^1(T^k)$ and have Fourier coefficients $c(n)$. Let

4.23 $$H(x) = C(k, \alpha) \int_{T^k} g_\alpha(x-y) f(y) dy .$$

As g_α and f are in L^1, $H(x)$ is in L^1. Also $H(x)$ has Fourier coefficients $h(n)$ where

$$h(n) = \begin{cases} 0 & \text{if } n = 0 \\ c(n)|n|^{-\alpha} & \text{if } n \neq 0. \end{cases}$$

Thus $H(x)$ is a function (and hence the unique function) in L^1 having Fourier coefficients that define $f_\alpha(x)$. (See 4.19). Thus $H(x) = f_\alpha(x)$ a.e., and $f_\alpha(x)$ is unique up to a set of measure zero. This finishes the proof of Theorem 18.

The idea of the proof of Theorem 19 is the following: we start with an $f(x)$ in $L^1(T^k)$. We first extend $f(x)$ to an $f_1(x)$ in $L^1(E_k)$, and we then apply Theorems A, B, and C to $f_1(x)$. However, we must be sure that $f_1(x)$ is in Lip α when $f(x)$ is in Lip α. This causes a certain amount of messiness in the proof. Another difficulty arises because the $g_\alpha(x)$ of Theorem 18 has a singularity not only at the origin, but at every lattice point.

Proof of Theorem 19. Let $f(x)$ be in $L^1(T^k)$, and assume $f(x)$ satisfies the hypothesis of Theorem A, B, or C. We shall write

$$f_\alpha(x) = h_1(x) + h_2(x).$$

$h_1(x)$ and $h_2(x)$ will not be periodic, but each will satisfy the conclusion of Theorem A, B, or C respectively in any compact subset of a certain region $\mathcal{S}$, where

$$\mathcal{S} = \{x \text{ in } E_k \mid -(3/2) < x_j < (3/2) \text{ for } j = 1, 2, \ldots k\} .$$

(By satisfying an integrability condition on a set, we mean that the integral over that set is finite.) It therefore will follow that $f_\alpha(x)$ satisfies the conclusions of Theorem A, B, or C respectively in T^k.

By 4.21 and the fact that g_α is periodic, we may write

$$g_\alpha(x) = C(k,\alpha) \sum_{\substack{n \\ 0 \le \max_j |n_j| \le 1}} |n+x|^{\alpha-k} + \mathscr{E}_\alpha(x),$$

where $\mathscr{E}_\alpha(x)$ is infinitely differentiable for all x such that

$$0 \le \max_j |x_j| < 2.$$

Therefore by Theorem 18,

$$f_\alpha(x) = h_1(x) + h_2(x).$$

$$h_1(x) = C(k,\alpha) \sum_{\substack{n \\ 0 \le \max_j |n_j| \le 1}} \int_{T^k} f(y) |n+x-y|^{\alpha-k} dy,$$

and

$$h_2(x) = \int_{T^k} f(y) \mathscr{E}_\alpha(x-y) dy.$$

Since $\mathscr{E}_\alpha(x)$ is infinitely differentiable for x satisfying $0 \le \max_j x_j < 2$, $h_2(x)$ is infinitely differentiable for x in $\mathscr{S}$ for f in $L^1(T^k)$. Hence it suffices to show that if $f(x)$ satisfies the hypothesis of Theorem A, B, or C respectively, then $h_1(x)$ satisfies the conclusion of Theorem A, B, or C respectively for x in $\mathscr{S}$.

We let $f_1(x)$ denote the periodic extension of f to all of E_k. That is $f_1(x)$ is defined by the conditions

$$f_1(x) = f(x)$$

for x in T^k, and

$$f_1(x+n) = f_1(x)$$

for any lattice point n. We also choose a function $U(x)$ in $C_0^\infty(E_k)$ such that $U(x) = 1$ on $\bar{\mathscr{S}}$ (the topological closure of $\mathscr{S}$). Now

$$h_1(x) = C(k,\alpha) \sum_{\substack{n \\ 0 \le \max |n_j| \le 1}} \int_{T^k} f_1(y)U(y)|n+x-y|^{\alpha-k} dy$$

$$= C(k,\alpha) \int_{\mathscr{S}} f_1(y)U(y)|x-y|^{\alpha-k} dy.$$

To establish the last equality we use the fact that f_1 is periodic and the fact that $U(x) = 1$ for x in $\mathscr{S}$. We can thus write

$$h_1(x) = h_3(x) - h_4(x);$$

where

$$h_3(x) = C(k,\alpha) \int_{E_k} f_1(y)U(y)|x-y|^{\alpha-k} dy,$$

and

$$h_4(x) = C(k,\alpha) \int_{E_k - \mathscr{S}} f_1(y)U(y)|x-y|^{\alpha-k} dy.$$

Let x_0 be a point in S. Then $x - y$ is bounded away from zero for y not in $\mathscr{S}$ and x in some neighborhood of x_0. Hence, it is clear that $h_4(x)$ is infinitely differentiable for x in $\mathscr{S}$.

Therefore it suffices to prove that if f satisfies the hypothesis of Theorem A, B, or C, then $h_3(x)$ satisfies the conclusion of Theorem A, B, or C respectively for x in $\mathscr{S}$. But if $f(x)$ satisfies the hypothesis of Theorem A, B, or C on T^k, then $f_1(y)U(y)$ satisfies the hypothesis of Theorem A, B, or C respectively on E_k. Therefore by using Theorem A, B, or C for fractional integration on E_k, we see that $h_3(x)$ satisfies the conclusion of Theorem A, B, or C respectively for x in E_k. A fortiori, $h_3(x)$ satisfies the conclusion of Theorem A, B, or C respectively for x in $\mathscr{S}$. This completes the proof of Theorem 19.

Proof of Theorem 20. We consider first fractional integration on E_k. In order to apply the theorems of part I to prove Theorem 20, we would like, for a given $f(x)$ in $L^1(E_k)$, to be able to characterize $f_\alpha(x)$ by a multiplier transformation. However we cannot quite do this, but we shall however show

4.24 $$f_\alpha(x) = g_\alpha(x) + E_1(x);$$

where

4.25 $$g_\alpha\hat{\ }(x) = \psi(|x|)|x|^{-\alpha} f\hat{\ }(x),$$

and $E_1(x)$ is infinitely differentiable. To prove 4.24 we begin by noting that

$$C(k,\alpha)|x|^{-\alpha} = F(x) - E(x)$$

where

$$F(x) = \lim_{\varepsilon \to 0+} \{\psi(|x|)|x|^{-\alpha} \exp(-\varepsilon|x|)\}\check{\ };$$

$E(x)$ is infinitely differentiable, and $E(x)$ and each of its derivatives is bounded on E_k. (The above observation follows from Remark 2 to Theorem 1.) Thus from the definition 4.14 we see that

$$f_\alpha(x) = g_\alpha(x) + E_1(x);$$

$E_1(x)$ is infinitely differentiable, and

$$g_\alpha(x) = \int_{E_k} F(x-y)f(y)dy.$$

By Theorem 1, $F(x)$ is in L^1 and

$$F\hat{\ }(x) = \psi(|x|)|x|^{-\alpha}.$$

Therefore by the convolution theorem $g_\alpha(x)$ is in L^1, and

$$g_\alpha\hat{\ }(x) = \psi(|x|)|x|^{-\alpha} f\hat{\ }(x).$$

This establishes 4.24 and 4.25.

We now prove assertion i). First $q > 1$ is clearly necessary in Theorem B, for otherwise α would be at least as large as k. We shall now assume $\alpha = k/q$. We must exhibit an $f(x)$ in $L^q(E_k) \cap L^1(E_k)$ such that $f_\alpha(x)$ is not in Lip 0. Take

$$f(x) = \lim_{\varepsilon \to 0+} \{\psi(|x|)|x|^{-k+k/q}(\log(|x|+2))^{-1} \exp(-\varepsilon|x|)\}\check{\ }.$$

It is clear from Theorem 1 that $f(x)$ exists and is in $L^q \cap L^1$. Also

$$f\hat{}(x) = \psi(|x|)|x|^{-k+k/q}(\log(|x|+2))^{-1}.$$

So by 4.24 and 4.25

$$\hat{g}_{k/q}(x) = (\psi(|x|))^2|x|^{-k}(\log(|x|+2))^{-1}$$

where $g_{k/q}$ is in L^1, and where $f_{k/q} - g_{k/q}$ is a bounded infinitely differentiable function. The Fourier inversion formula implies

$$g_{k/q}(x) = \lim_{\varepsilon \to 0+} \{(\psi(|x|))^2|x|^{-k}(\log(2+|x|))^{-1}\exp(-\varepsilon|x|)\}^{\vee}.$$

As $\int_2^\infty (t\log t)^{-1}dt = \infty$, Theorem 4 implies $g_{k/q}$ is not locally bounded, hence $f_{k/q}$ is not locally bounded, which is what we wanted to prove.

We now set $\alpha = 1 + k/q$, and we wish to find an $f(x)$ in $L^1 \cap L^q$ such that $f(x)$ is not in Lip 1. We consider

$$f(x) = \lim_{\varepsilon \to 0+} \{\psi(|x|)x_1|x|^{-1-k+k/q}(\log(|x|+2))^{-1}\exp(-\varepsilon|x|)\}^{\vee}.$$

By Theorem 1 of part I, it is clear that $f(x)$ is in $L^1 \cap L^q$. (Here $Y_\ell(x') = Y_1(x') = x_1(|x|)^{-1}$.) Also

$$f\hat{}(x) = \psi(|x|)x_1|x|^{-1-k+k/q}(\log(|x|+2))^{-1}.$$

So by 4.24 and 4.25

$$g_{1+k/q}\hat{}(x) = x_1|x|^{-2-k}(\psi(|x|))^2(\log(|x|+2))^{-1}$$

where $g_{1+k/q}$ is in L^1, and $g_{1+k/q}$ differs from $f_{1+k/q}$ by an infinitely differentiable function. Now by the Fourier inversion formula

$$g_{1+k/q}(x) = \{(\psi(|x|))^2x_1|x|^{-2-k}(\log(|x|+2))^{-1}\}^{\vee}.$$

By Theorem 5,

$$\frac{\partial}{\partial x_1}g_{1+k/q}(x) = \lim_{\varepsilon \to 0+} \{(\psi(|x|))^2x_1^2|x|^{-2-k}(\log(|x|+2))^{-1}\exp(-\varepsilon|x|)\}^{\vee}.$$

Now

$$x_1^2 |x|^{-2} = C + \sum_{\substack{\ell, m \\ \ell > 0}} a(\ell, m) Y_{\ell, m}(x'),$$

where the range of summation is finite and $C \neq 0$ since

$$\int_{S_{k-1}} x_1^2 |x|^{-2} dx' \neq 0.$$

(See [6] chapter 6, p. 31, Proposition 3(b) and the definition above it.) We thus have

$$\frac{\partial}{\partial x_1} g_{1+k/q}(x) = v_1(x) + v_2(x).$$

$$v_1(x) = C \lim_{\varepsilon \to 0+} \{(\psi(|x|))^2 |x|^{-k} (\log(|x| + 2))^{-1} \exp(-\varepsilon |x|)\}^{\vee},$$

and

$$v_2(x) = \sum_{\substack{\ell, m \\ \ell > 0}} a(\ell, m) \{(\psi(|x|))^2 |x|^{-k} (\log(|x|+2))^{-1} Y_\ell(x') \exp(-\varepsilon |x|)\}^{\vee},$$

where the sum in the above expression for v_2 is finite.

By Theorem 4, $h_1(x)$ is not locally bounded, and $h_2(x)$ is bounded. Hence $\frac{\partial}{\partial x_1} g_{1+k/q}(x)$ is not locally bounded. Hence $\frac{\partial}{\partial x_1} f_{1+k/q}$ is not locally bounded, and this is what we wanted to prove.

We are now finished with i) and turn to the proof of ii). The fact that $\alpha < k/q$ cannot be improved to $\alpha = k/q$ in Theorem C, is the same as the second of the three assertions of i). To prove the necessity of the condition $q > 1$ in Theorem C, take

$$f(x) = \lim_{\varepsilon \to 0+} \{\psi(|x|)(\log(|x|+2))^{(\alpha-k)/k}\}^{\vee}.$$

By Theorem 3, $f(x)$ is in L^1, and

$$f^{\wedge}(x) = \psi(|x|)(\log(|x|+2))^{(\alpha-k)/k}.$$

So by 4.24 and 4.25

$$f_\alpha(x) = g_\alpha(x) + E_1(x);$$

where

$$g_\alpha\hat{\ }(x) = (\psi(|x|))^2 |x|^{-\alpha}(\log(|x|+2))^{(\alpha-k)/k};$$

$g_\alpha(x)$ is in L^1, and $E_1(x)$ is infinitely differentiable. By the Fourier inversion formula

$$g_\alpha(x) = \lim_{\varepsilon \to 0+} \{\psi^2(|x|)|x|^{-\alpha}(\log(|x|+2))^{(\alpha-k)/k}\exp(-\varepsilon|x|)\}^{\vee}.$$

By Theorem 1,

$$g_\alpha(x) \simeq C(k,\alpha)|x|^{\alpha-k}(\log(1/|x|))^{(\alpha-k)/k},$$

as $x \to 0$. Therefore $g_\alpha(x)$ is not locally in L^q for $q = k/(k-\alpha)$. Hence the same is true for $f_\alpha(x)$. This completes the proof of Theorem 20 for fractional integration on E_k. The proof of Theorem 20 for fractional integration on T^k is similar. We use the theorems of part I concerning series rather than the theorems concerning integrals. We omit the details. This finishes our discussion of fractional integration.

We turn now to the study of Riesz transforms. Let $f(x)$ be in $L^1(E_k)$. We let

4.26 $$\mathscr{R}_j f(x) = \lim_{\varepsilon \to 0+} \int_{|x-y|>\varepsilon} (x_j - y_j)|x-y|^{-k-1} f(y)dy$$

for $j = 1, 2, \ldots, k$. (See [7]).

If $f(x)$ is in $L^1(T^k)$, we set

4.27 $$\mathscr{R}_j f(x) = \lim_{\varepsilon \to 0+} \int_{\substack{|x-y|>\varepsilon \\ y \text{ in } T^k}} K_j(x-y)f(y)dy$$

where

$$K_j(x) = x_j|x|^{-k-1} + \sum{}' \{(x_j+n_j)|x+n|^{-k-1} - n_j|n|^{-k-1}\}.$$

(See [8].) In either 4.26 or 4.27, we call $\mathscr{R}_j f$ the jth Riesz transform of f.

It is known that if $f(x)$ is in L^1 of E_k (or T^k), then $\mathscr{R}_j f$ exists for almost every x if the limit is taken in the pointwise sense. Furthermore if $f(x)$ is in L^p, $p > 1$, then $\mathcal{R}_j f$ is in L^p, and if $f(x)\log(1 + |f(x)|)$ is in L^1, then $\mathcal{R}_j f$ is in L^1. (See [7] and [8]).

Our object here will be to give a simple example of a function $f(x)$ such that

$$\int_{E_k} |f(x)| \log^{1-\delta}(1 + |f(x)|)dx$$

is finite for every $\delta > 0$ and such that $\mathcal{R}_j f$ is not in L^1 (for every j). Or to be slightly more general, we shall prove

THEOREM 21. Let $\beta \geq 0$.

i) Let

$$F(x) = \lim_{\varepsilon \to 0+} \{\psi(|x|)\log^{-\beta-1}(|x|+2)\exp(-\varepsilon|x|)\}^{\vee}.$$

Then for any $\delta > 0$,

$$\int_{E_k} |F(x)| \log^{\beta+1-\delta}(1 + |F(x)|)dx < \infty,$$

and

$$\int_{E_k} |\mathcal{R}_j F| \log^{\beta}(1 + |\mathcal{R}_j F|)dx = \infty .$$

ii) Let

$$f(x) = \lim_{\varepsilon \to 0+} \sum \log^{-\beta-1}(|n|+2)\exp(-\varepsilon|n| + 2\pi i n \cdot x).$$

Then for any $\delta > 0$

$$\int_{T^k} |f(x)| \log^{\beta+1-\delta}(1 + |f(x)|)dx < \infty,$$

and

$$\int_{T^k} |\mathcal{R}_j f| \log^{\beta}(1 + |\mathcal{R}_j f|)dx = \infty .$$

<u>Proof of i)</u>. Formally one obtains that for $f(x)$ in L^1,

4.28 $$(\mathscr{R}_j f)^\wedge = (x_j|x|^{-k-1})^\wedge f^\wedge$$

$$= C(j,k)x_j|x|^{-1}f^\wedge,$$

with $C(j,k) \neq 0$. It then would be an easy matter to use the theorems of part I to prove i). Therefore we shall first prove a lemma essentially equivalent to 4.28.

LEMMA 20. Let $g(x)$ be in $L^1(E_k)$. Define,

$$\mathscr{H}_{j,\varepsilon}g = \{x_j|x|^{-1}\exp(-\varepsilon|x|)g^\wedge(x)\}^\vee.$$

Then for almost every x

4.29 $$\mathscr{R}_j g = C(j,k)\lim_{\varepsilon\to 0+}\mathscr{H}_{j,\varepsilon}g.$$

(The limit as $\varepsilon \to 0+$ in equation 4.29 is to be interpreted as being in the point-wise sense.)

Proof of Lemma 20. Horvath shows that for almost every x

4.30 $$\lim_{\varepsilon\to 0+}\left\{\int_{|x-y|>\varepsilon}(y_j-x_j)|y-x|^{-k-1}g(y)dy - \right.$$

$$\left. - \int_{E_k}(x_j-y_j)(|x-y|^2+\varepsilon^2)^{-\frac{1}{2}(k+1)}g(y)dy\right\} = 0.$$

(See [15]). He also shows

4.31 $$\{x_j|x|^{-1}\exp(-\varepsilon|x|)\}^\vee = C(j,k)(x^2+\varepsilon^2)^{-\frac{1}{2}(k+1)}$$

for $\varepsilon > 0$ (with $C(j,k) \neq 0$). Since g is in L^1, and

$$x_j(x^2+\varepsilon^2)^{-\frac{1}{2}(k+1)}$$

is in L^2,

$$\int_{E_k} g(y)(x_j-y_j)(|x-y|^2+\varepsilon^2)^{-\frac{1}{2}(k+1)}dy$$

is in L^2. So by the convolution Theorem

4.32 $$\left\{\int g(y)(x_j-y_j)(|x-y|^2+\varepsilon^2)^{-\frac{1}{2}(k+1)}dy\right\}^{\wedge 2}$$

$$= g^\wedge\{x_j(x^2+\varepsilon^2)^{-\frac{1}{2}(k+1)}\}^{\wedge 2}$$

$$= g^\wedge(C(j,k))^{-1}x_j|x|^{-1}\exp(-\varepsilon|x|)$$

by 4.31. (By $^{\wedge 2}$ we mean the Fourier transform in the L^2 sense.)

For functions in $L^1 \cap L^2$, the inverse Fourier transforms taken in the ordinary sense and in the L^2 sense are the same. So the Fourier inversion formula for L^2 together with 4.32 gives

$$\mathscr{H}_{j,\varepsilon}g = \int_{E_k} g(y)(x_j-y_j)(|x-y|^2+\varepsilon^2)^{-\frac{1}{2}(k+1)}dy.$$

We are now finished with the proof of Lemma 20 because of 4.30.

We shall now complete the proof of i). By Theorem 3, $F(x)$ exists and is continuous for $x \neq 0$. Also as $x \to 0$

$$F(x) \simeq K(k,\beta)|x|^{-k}(\log(1/|x|))^{-\beta-2}.$$

So clearly,

$$\int |F(x)|\log^{1+\beta-\delta}(1+|F(x)|)dx < \infty.$$

Also

$$F^\wedge(x) = \psi(|x|)\log^{-\beta-1}(2+|x|).$$

By Lemma 20,

$$\mathscr{R}_jF = \lim_{\varepsilon\to 0+} H_{j,\varepsilon}F$$

$$= \lim_{\varepsilon\to 0+} C(j,k)\{\psi(|x|)x_j|x|^{-1}\log^{-\beta-1}(|x|+2)\exp(-\varepsilon|x|)\}^\vee.$$

By Theorem 1, then,

$$\mathscr{R}_jF = K(k,\beta)x_j|x|^{-k-1}\log^{-\beta-1}(1/|x|) + o\{|x|^{-k}\log^{-\beta-1}(1/|x|)\}$$

as $x \to 0$. Therefore,

$$\int_{E_k} |\mathscr{R}_j F| \log^\beta (1 + |\mathscr{R}_j F|) dx = \infty .$$

This finishes the proof of i).

The proof of ii) is similar to the proof of i), and we omit the details. We only remark that the appropriate analogue of Lemma 20 appears in the literature (see [8]).

APPENDIX

The purpose of this appendix is to give a large class of slowly varying functions which are in the class S. (See def. 2 of part I.) We need to recall a definition due to Hardy. (See [12], p. 17).

"We define a LOGARITHMICO-EXPONENTIAL FUNCTION (shortly an L-FUNCTION) as a real one-valued function defined, for all values of X greater than some definite value, by a finite combination of the ordinary algebraic symbols ($+$, $-$, $\cdot$, $\div$, ^{n}, and $\sqrt[n]{\ }$) and the functional symbols $\log(\ldots)$ and $e^{(\ldots)}$, operating on the variable X and on real constants."

Hardy continues in small print:

"It is to be observed that the result of working out the value of the function, by substituting X in the formula defining it, is to be real at all stages of the work. It is important to exclude such a function as

$$\tfrac{1}{2}\,\{\exp(-X^2)^{\frac{1}{2}} + \exp(-(-X^2)^{\frac{1}{2}})\}$$

which, with a suitable interpretation of the roots, is equal to $\cos X$."

REMARK: If a slowly varying function $b(t)$ is an infinitely differentiable L-function, then $b(t)$ is in the class S.

The remark follows from the following two facts about L-functions (see [12], p. 18).

i) the derivative of an L-function is an L-function.

ii) any L-function is eventually of constant sign and monotonic.

BIBLIOGRAPHY

1. Berg, L. Asymptotische Darstellungen fur Verallgemeinerte Fourierintegrale, Mathematische Nachrichten, 20, 1959, pp. 166-170.

2. __________. Uber das Asymptotische Verhalten der Inversen Laplace-Transformationen, Mathematische Nachrichten, 22, 1960, pp. 87-91.

3. Bernstein, S. Sur la Convergence absolue des series trigonometrique, Comm. Soc. Math. Kharkov, 2 me serie 14 (1914), pp. 139-144.

4. Bochner, S. Harmonic Analysis and the Theory of Probability. Berkeley and Los Angeles, 1955, University of California Press.

5. __________. Review of "On Absolute Convergence of Multiple Fourier Series" by Szasz and Minakshisundaram, Mathematical Reviews, 8, 1947, p. 376.

6. Calderón, A. P. Integrals Singulares y sus Applicaciones a Equaciones Diferenciales Hiperbolicas, 1960, Buenos Aires, University of Buenos Aires.

7. Calderón, A. P. and Zygmund, A. On the Existence of Certain Integrales, Acta Mathematica, 88, 1952, pp. 85-139.

8. __________. Singular Integrals and Periodic Functions, Studia Mathematica, 14, 1953-54, pp. 249-271.

9. De Bruijn, N. Asymptotic Methods in Analysis. Amsterdam, 1958, North Holland Publishing Co.

10. Du Plessis, N. Some Theorems about the Riesz Fractional Integral, Trans. of the Amer. Math. Soc., 80, 1955, pp. 124-134.

11. __________. Spherical Fractional Integrals, Trans. of the Amer. Math. Soc., 84, 1957, pp. 262-272.

12. Hardy, G. H. Orders of Infinity. Cambridge (1924), Cambridge University Press.

13. __________. A Theorem Concerning Taylor's Series, Quarterly Journal of Pure and Applied Mathematics, 44, 1913, pp. 147-160.

14. Hirschman, I. I. On Multiplier Transformations, Duke Mathematical Journal, 26, 1959, pp. 221-242.

15. Horvath, J. Sur Les Fonctions Conjugees A Plusieurs Variables, Proceedings Akademie Van Wetenshappen, Series A, 56, 1953, pp. 17-29.

16. Littlewood, J. E. Theory of Functions, Oxford, 1944, Oxford University Press.

17. Marcinkiewicz, J. Sur les Multiplicateurs des Series de Fourier, Studia Mathematica, 8, 1939, pp. 78-91.

18. Mikusinski, J. On the Function whose Laplace Transform is e^{-s^a}, Studia Mathematica, 18, 1959, pp. 191-198.

19. Rudin, W. Factorization in Group Algebras, American Mathematical Society Bulletin, 63, 1953, p. 382.

20. Stein, E. M. Interpolation of Linear Operators, Trans. of the American Math. Soc., 83, 1956, pp. 482-492.

21. Szasz, O. and Minakshisundaram, S. On Absolute Convergence of Fourier Series, Trans. of the American Math. Soc., 61, 1947, pp. 36-53.

22. Thorin, G. O. Convexity Theorems, Meddelanden Fran, Lunds Universitets Matematiska Seminarium, 9, 1948, pp. 1-57.

23. Titchmarsh, E. C. Fourier Integrals, 1937. Oxford, Oxford University Press.

24. Watson, G. N. Theory of Bessel Functions, 1944, Cambridge, Cambridge University Press.

25. Zygmund, A. Trigonometric Series. 1959, Cambridge, Cambridge University Press.

University of Chicago,
DePaul University, Chicago
and
Cornell University, Ithaca, N. Y.